Advances in the Physics of Particles and Nuclei
Volume 30

Advances in the Physics of Particles and Nuclei

The series *Advances in the Physics of Particles and Nuclei* (APPN) is devoted to the archiving, in printed high-quality book format, of the comprehensive, long shelf-life reviews published in *The European Physical Journal A* and *C*. APPN will be of benefit in particular to those librarians and research groups, who have chosen to have only electronic access to these journals. Occasionally, original material in review format and refereed by the series' editorial board will also be included.

Advances in the Physics of Particles and Nuclei

Volume 30

Edited by

Douglas H. Beck
Dieter Haidt
John W. Negele

Volume 30

Contributions to this Volume:

QCD Thermodynamics from the Lattice

Hadronic Parity Violation and Effective Field Theory

Cosmic Microwave Background and First Molecules
in the Early Universe

The 2009 World Average of α_s

 Springer

Douglas H. Beck
Department of Physics
University of Illinois at Urbana-Champaign
1110 West Green Street
Urbana, IL 61801-3080
USA
e-mail: dhbeck@illinois.edu

Dieter Haidt
DESY
Notkestraße 85
22603 Hamburg
Germany
e-mail: dieter.haidt@desy.de

John W. Negele
William A. Coolidge Professor of Physics
Massachusetts Institute of Technology
Center for Theoretical Physics
77 Massachusetts Ave. 6-315
Cambridge MA 02139
USA
e-mail: negele@MIT.EDU

Originally published in Eur. Phys. J. A 41, 405–437 (2009), Eur. Phys. J. A 41, 279–298 (2009),
Eur. Phys. J. C 59, 117–172 (2009), and Eur. Phys. J. C 64, 689–703 (2009)
© Springer-Verlag / Società Italiana di Fisica 2009

ISSN 1868-2146   e-ISSN 1861-440X
ISBN 978-3-662-56846-0   ISBN 978-3-642-04123-5   (eBook)
DOI 10.1007/978-3-642-04123-5
Springer Heidelberg Dordrecht London New York

Cover design: eStudioCalamar, Figueres

Printed on acid-free paper

Springer is part of Springer Science+Business Media (www.springer.com)

Table of Contents

QCD thermodynamics from the lattice

C.E. DeTar[1,a] and U.M. Heller[2,b]

[1] Physics Department, University of Utah, Salt Lake City, UT 84112-0830, USA
[2] American Physical Society, One Research Road, Ridge, NY 11961, USA

Received: 11 May 2009
Published online: 19 July 2009 – © Società Italiana di Fisica / Springer-Verlag 2009
Communicated by U.-G. Meißner

Abstract. We review the current methods and results of lattice simulations of quantum chromodynamics at nonzero temperatures and densities. The review is intended to introduce the subject to interested nonspecialists and beginners. It includes a brief overview of lattice gauge theory, a discussion of the determination of the crossover temperature, the QCD phase diagram at zero and nonzero densities, the equation of state, some in-medium properties of hadrons including charmonium, and some plasma transport coefficients.

PACS. 12.38.Gc Lattice QCD calculations – 12.38.Mh Quark-gluon plasma – 21.65.Qr Quark matter – 25.75.Nq Quark deconfinement, quark-gluon plasma production, and phase transitions

Contents

a e-mail: detar@physics.utah.edu
b e-mail: heller@aps.org

1 Introduction

Quantum chromodynamics is the well-established theory of interacting quarks and gluons. Although its Lagrangian is simple and elegant, except for high-energy processes where perturbation theory is applicable, it is very difficult to solve. Over the past three decades *ab initio* numerical and computational methods have been devised for obtaining nonperturbative solutions. They have become refined to the point that a few dozen calculated quantities (decay constants, mass splittings, etc.) agree with known experimental values to a precision of a couple percent [1]. These successes provide the opportunity to push the calculations with some confidence into new regimes that have not been thoroughly explored experimentally. In this review we will be interested in numerical simulations of strongly interacting matter under the extreme conditions of high temperatures and/or high baryon number densities.

Shortly after the big bang the Universe was very likely dominated by a high-temperature plasma of quarks, antiquarks, and gluons. As the Universe expanded and cooled, hadrons emerged that make up today's Universe. Knowing the characteristics of the plasma and the nature of the transition to hadrons is clearly important for understanding these stages in the development of the Universe. In the cores of some dense stars it is conceivable that the baryon number density is sufficiently high that hadrons lose their identities and merge into a plasma of quarks and gluons. The equation of state of such a dense plasma, for example, is important for understanding conditions leading to a collapse to a black hole. In heavy-ion collisions at RHIC, FAIR, and soon at the LHC we seek to produce a quark-gluon plasma and study its properties. Since so little is known about the plasma, we turn to numerical simulation of high-temperature and moderate-density QCD to predict its properties and to guide the experimental investigation. Apart from the phenomenological interest in such simulations, there is also intrinsic theoretical interest in understanding the behavior of confining field theories under extreme conditions. In particular, there are tantalizing predictions of still new states of matter at very high densities [2]. Lattice QCD thermodynamics is understandably a popular and vigorous field of research.

Certainly, present-day lattice simulations cannot answer all of our questions. The current standard methodology assumes thermal equilibrium. Moreover, simulations at nonzero densities are still in their infancy, so much of what we know is restricted to zero or very small baryon number densities. To apply lattice results to the phenomenology of heavy-ion collisions requires an intermediate model, such as hydrodynamics, which takes input from lattice simulations, adds model assumptions, and makes predictions about the rapidly evolving, emerging matter. For this purpose the most important quantities obtained from lattice simulations are the phase diagram as a function of temperature and baryon number density, the equation of state, speed of sound, and transport properties, such as the viscosities and thermal conductivity.

In this review, intended for nonspecialists and beginners, we give a brief overview of the lattice methodology and discuss a variety of numerical results. We discuss challenges and potential sources of systematic error. In sect. 2 we give a brief introduction to lattice gauge theory and discuss the advantages and disadvantages of various fermion formulations. We discuss a variety of observables used to determine the transition temperature T_c in sect. 3 and comment on some disparate results. In sect. 4 we review our current understanding of the phase diagram at zero baryon density, and in sect. 5 we do the same for nonzero baryon number densities. We discuss the variety of methods in current use for simulating at nonzero densities. We review the equation of state in sect. 6. In sect. 7 we discuss some properties of hadrons in the high-temperature medium, and in sect. 8 some results for transport coefficients. Finally, in sect. 9 we summarize briefly the current state of the field, list outstanding problems, and list some prospects for resolving them.

2 Thermodynamics in lattice gauge theory

2.1 Quantum partition function

Quantum thermodynamics at a fixed, large volume is based on the partition function in the quantum canonical ensemble

$$Z = \text{Tr}[\exp(-H/T)], \qquad (1)$$

where H is the quantum Hamiltonian operator, T is the temperature, and the trace is taken over the physical Hilbert space. At nonzero densities the grand canonical ensemble is appropriate:

$$Z = \text{Tr}\left[\exp\left(-H/T + \sum_i \mu_i N_i/T\right)\right], \qquad (2)$$

where μ_i is the chemical potential for the i-th species and N_i is the corresponding conserved flavor number. For example, in QCD we may introduce a separate chemical potential for each quark flavor. Zero chemical potential for a given flavor implies equal numbers of quarks and antiquarks of that flavor, so zero baryon number density, zero strangeness, etc.

The expectation value of an observable $\mathcal{O}$ at temperature T is computed with respect to this ensemble through

$$\langle \mathcal{O} \rangle = \text{Tr}\left[\mathcal{O} \exp\left(-H/T + \sum_i \mu_i N_i/T\right)\right]\Big/ Z. \qquad (3)$$

2.2 Feynman path integral partition function

The Feynman path integral formalism provides a practical basis for the computation of thermodynamic quantities, especially in quantum field theory, where there are many degrees of freedom. It converts the trace over quantum states into a multidimensional integration over classical variables [3]. It is beyond the scope of this review to give a detailed derivation of the path integral formulation, particularly for a gauge theory with fermion fields. There are standard references [4–6].

The classical variables in the Feynman path integral are the path "histories" of the fundamental fields in Euclidean (imaginary) time τ. (Imaginary, because the Boltzmann weight factor $\exp(-H/T)$ is, in effect, a time evolution operator $\exp(-iHt)$ for an imaginary time $-i/T$.) For computational purposes the histories are discretized in τ. The quantum fields at any given time are also discretized in three-dimensional coordinate space $\mathbf{x}$. The resulting path integral is then a multidimensional integral over variables defined on a four-dimensional space-time lattice ($x = \mathbf{x}, \tau$). The discretization of space and time introduces an error, but the error vanishes as the lattice spacing is taken to zero (continuum limit).

2.3 Scalar field example

For a concrete example, consider a scalar field theory described by the Lagrange density

$$L(\phi) = \frac{1}{2} \sum_{\mu} \left[\frac{\partial}{\partial x_{\mu}} \phi(x) \right]^2 + V[\phi(x)], \qquad (4)$$

where V describes the mass and self-interaction. On a hypercubic lattice with point separation a and a central-difference discretization of the derivatives, we can write a lattice approximation

$$L[\phi(x)] = \frac{1}{8a^2} \sum_{\mu} \left[\phi(x + a\hat{\mu}) - \phi(x - a\hat{\mu}) \right]^2 + V[\phi(x)]. \qquad (5)$$

where $\hat{\mu}$ is a unit coordinate vector in the μ direction. A Euclidean time history is then specified simply by giving the values of the field $\phi(\mathbf{x}, \tau)$ on all the lattice points $(\mathbf{x}, \tau)$. Each such history corresponds to a classical Euclidean action $S(\phi)$, which is computed by summing its Lagrange density over the lattice points

$$S(\phi) = \sum_{\mathbf{x}, \tau} L[\phi(\mathbf{x}, \tau)]. \qquad (6)$$

The partition function then becomes a multidimensional integral over the values of the field $\phi(\mathbf{x}, \tau)$ at each point, weighted by the exponential of the classical Euclidean action:

$$Z = \int \prod_{\mathbf{x}, \tau} d\phi(\mathbf{x}, \tau) \exp[-S(\phi)]. \qquad (7)$$

Two important conditions on the Euclidean time history are inherited from the definition (eq. (1)) of the partition function: First, the time history τ ranges over a finite interval from 0 to $a(N_{\tau} - 1)$ where

$$1/T = aN_{\tau}, \qquad (8)$$

which establishes the relation between the temperature and the Euclidean time extent of the lattice. Second, to reproduce the trace over quantum states, the bosonic field ϕ must be periodic under $\tau \to \tau + aN_{\tau}$.

Similarly, the expectation value of an operator $\mathcal{O}(\phi)$, which depends on the field ϕ, is given by

$$\langle \mathcal{O} \rangle = \int \prod_{\mathbf{x}, \tau} d\phi(\mathbf{x}, \tau) \, \mathcal{O}(\phi) \exp[-S(\phi)]/Z, \qquad (9)$$

where we replace the field operator ϕ with its classical value when we insert $\mathcal{O}(\phi)$ in the integrand.

2.4 QCD on the lattice

For a renormalizable, asymptotically free theory, such as QCD, the lattice formulation takes on a larger significance than just a convenient computational device. The lattice regulates ultraviolet divergences. The lattice constant a provides an upper bound or "cutoff" scale π/a for momenta. From this point of view the lattice formulation of the theory is every bit as respectable as other regularization schemes. Of course, as usual, we define the theory in the limit in which the cutoff is removed, i.e., $a \to 0$. Before this is done all quantities we calculate have cutoff errors that vanish in the continuum limit.

With the lattice regulator we apply the usual renormalization process: we select a few experimental values and use them to fix the bare parameters of the theory (quark masses and gauge coupling). In this way the bare parameters depend on the cutoff (lattice spacing), but the physical predictions should approach a cutoff-independent value in the limit of zero lattice spacing. In principle all regularization schemes should agree in the limit that their cutoffs are taken away.

For QCD the fields are fermions and gluons. The groundwork for the lattice formulation of QCD with fermions was laid down by Wilson [7] in 1974, although there was other seminal work on lattice theories with local gauge invariance by Wegner [8], Smit, and Polyakov [9]. To preserve gauge invariance, gluon variables are introduced as $SU(3)$ matrices on the links between nearest neighbors of the lattice. There are four forward links per site, corresponding to the four components of the color vector potential $A_{\mu}^c(x)$. The matrix for the link joining the site x with the site $x + a\hat{\mu}$ is then

$$U_{\mu}(x) = \exp[igaA_{\mu}^c(x)\lambda^c/2], \qquad (10)$$

where λ^c are the Gell-Mann generators of $SU(3)$.

For the pure Yang-Mills theory of gluons a simple lattice form of the classical action is constructed from the plaquettes $U_{P,\mu\nu}(x)$, i.e., the product of the link matrices surrounding the unit square in the forward $\mu\nu$ direction at site x. The single-plaquette Wilson action is simply the sum over all such plaquettes:

$$S_G(U) = \sum_{x,\mu,\nu} \frac{\beta}{6} \operatorname{Re} \operatorname{Tr}[1 - U_{P,\mu\nu}(x)], \qquad (11)$$

The gauge coupling $\alpha_s = g^2/4\pi$ appears in the coefficient

$$\beta = 6/g^2. \qquad (12)$$

In the continuum limit the plaquette reduces to the familiar square of the field strength tensor summed over eight colors c:

$$\mathrm{Re\,Tr}[1 - U_{P,\mu\nu}(x)] \to \frac{g^2 a^4}{4} \sum_c (F^c_{\mu\nu})^2 + \mathcal{O}(a^6). \quad (13)$$

In fact any closed planar loop, normalized by the area in lattice units, has the same continuum limit, but with a different $\mathcal{O}(a^6)$ cutoff error. For example a 2×1 rectangular version of the plaquette could also be used. If the two components are combined with the proper choice of coefficients, one can construct an improved gluon action that eliminates the leading cutoff correction, leaving errors at the next order $\mathcal{O}(a^8)$. Relative to the leading continuum contribution, which carries the volume factor a^4, such actions are called "tree level $\mathcal{O}(a^2)$" improved. Further improvements can even eliminate quantum cutoff corrections of the type $\mathcal{O}(a^2 \alpha_s)$. The "tadpole Lüscher-Weisz" actions [10,11] are in this category. Improving actions in this way is desirable, since it brings a calculation closer to the continuum limit at a given lattice spacing [12].

The quark fields $\psi(x)$, one for each flavor, have values on each lattice site. Since they are fermions, they require special treatment in the functional integration: their classical values are anticommuting Grassmann numbers. The fermion contribution to the action for each flavor can be written generally as

$$S_F(U, \psi) = \sum_{x,y} \bar{\psi}(x) M(U, x, y) \psi(y), \quad (14)$$

where $M(U, x, y)$ is the Dirac matrix —essentially a lattice rendering of the familiar Dirac operator $D\!\!\!\!/ + m$. The functional integral for the partition function then becomes

$$Z = \int [\mathrm{d}U][\mathrm{d}\psi][\mathrm{d}\bar{\psi}] \exp[-S_G(U) - S_F(U, \psi)]. \quad (15)$$

Since the dependence on the quark fields is simply bilinear, and computing numerically with anticommuting numbers is nontrivial, it is standard to integrate out the quark fields immediately, following the rules of Grassmann integration, leaving only an integration over the gauge fields, weighted by the determinant of the Dirac matrix:

$$Z = \int [\mathrm{d}U] \exp[-S_G(U)] \det[M(U)]. \quad (16)$$

There are many ways to formulate a lattice fermion action, each with its advantages and disadvantages. A great deal of effort over the past couple of decades has been devoted to improving the lattice treatment of fermions. We sketch the formulations here. For more detail, see [4–6].

2.4.1 Wilson fermions

The original Wilson rendering of the Dirac operator $D_\nu \gamma_\nu + m$ starts from a simple central-difference approximation to the derivative:

$$\nabla_\nu \psi(x) = \frac{1}{2a} [U_\nu(x) \psi(x + \hat{\nu}a) - U_\nu^\dagger(x - \hat{\nu}a) \psi(x - \hat{\nu}a)], \quad (17)$$

where the link matrices $U_\nu(x)$ provide the gauge covariance. The action constructed from this operator is

$$S_{F,\mathrm{naive}} = \sum_{x,\nu} \bar{\psi}(x)(\nabla_\nu \gamma_\nu + m)\psi(x). \quad (18)$$

It describes sixteen degenerate particles where only one is desired. Wilson remedied this undesirable "fermion doubling" problem by adding an irrelevant term to the action

$$S_{F,W} = S_{F,\mathrm{naive}} - \frac{ar}{2} \sum_{x,\nu} \bar{\psi}(x) \Delta_\nu \psi(x), \quad (19)$$

where r is usually set to 1 and $\Delta_\nu \psi(x)$ is the covariant Laplacian,

$$\Delta_\nu \psi(x) = \frac{1}{a^2} [U_\nu(x)\psi(x+\hat{\nu}a) + U_\nu^\dagger(x-\hat{\nu}a)\psi(x-\hat{\nu}a) - 2\psi(x)]. \quad (20)$$

The added term gives fifteen of the doublers masses of order of the cutoff scale $1/a$, leaving only one light state. The unwanted doublers thus become inaccessibly heavy in the continuum limit.

It is customary to rearrange the terms in the Wilson action and multiply the field $\psi(x)$ by a constant to give

$$S_{F,W} = \bar{\psi}(x)\psi(x) + \kappa \sum_{x,\nu} \bar{\psi}(x) \left[(1+\gamma_\nu) U_\nu(x)\psi(x + \hat{\nu}a) \right.$$
$$\left. + (1 - \gamma_\nu) U_\nu^\dagger(x - \hat{\nu}a)\psi(x - \hat{\nu}a) \right], \quad (21)$$

where the "hopping parameter" $\kappa = 1/(8 + 2ma)$ controls the quark mass. Improvements to the Wilson formalism include removing tree level $\mathcal{O}(a)$ errors by introducing a "clover" term in the action [13] and, for two flavors, introducing a "twisted mass" [14,15].

For thermodynamics applications the chief drawback of Wilson fermions has been 1) an explicit breaking of chiral symmetry at nonzero lattice spacing, 2) a difficulty reaching low quark masses, and 3) a relatively poor representation of the quark dispersion relation. None of these difficulties is insurmountable. Chiral symmetry is restored in the continuum limit.

It is necessary to search for the value $\kappa = \kappa_c$ where the pion mass vanishes. Since this value depends on the inverse gauge coupling β, one gets a curve $\kappa_c(\beta)$ in the bare parameter κ-β space as shown in fig. 1. Lines of constant pion mass form a family of such curves (not shown) with the pion mass increasing as κ decreases. Also shown is a high-temperature crossover line $\kappa_t(\beta)$. Its location depends on N_τ. Where it intersects the κ_c line, we expect a true chiral phase transition. Pushing to stronger coupling (smaller β) or negative quark masses (higher κ) from there takes us into the realm of lattice artifacts: the theory has a parity-broken phase at unphysical values of the bare parameters, as indicated.

2.4.2 Staggered fermions

The staggered fermion approach starts from the naive action in eq. (18). Through a field transformation, the Dirac

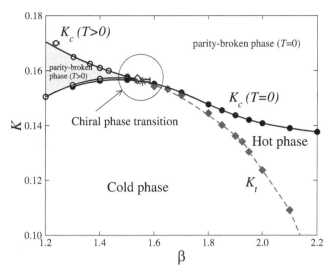

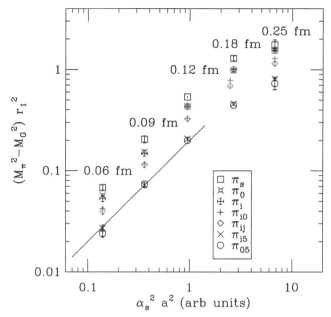

Fig. 1. The bare parameter phase diagram for two flavors of clover-improved Wilson fermions and an improved gauge action for zero and nonzero temperatures, illustrating the mapping necessary for thermodynamics studies with Wilson fermions. In this plot the hopping parameter κ is denoted by K. The line of chiral critical hopping parameters $\kappa_c(T = 0)$ was determined from the vanishing of the pion mass. The line κ_t indicates the high-temperature crossover at $N_\tau = 4$. It was determined from the Polyakov line (see sect. 3.1). The region "chiral phase transition" shows where the thermal crossover happens for small pion masses. The parity-broken phases come from lattice artifacts of Wilson fermions. The data are from the CP-PACS Collaboration [16], as shown in [17].

Fig. 2. Plot showing that the lattice artifact taste splitting of pion masses vanishes as $\alpha_s^2 a^2$ in the continuum limit. The splitting is measured as the difference of the squared masses of the multiplet member and the Goldstone pion member. It is given in units of $r_1 \approx 0.318\,\text{fm}$. The plot symbols distinguish the members of the multiplet. (The subscripts in the legend denote the Dirac-gamma-matrix-style classification of the pion tastes, ranging from singlet (s) and γ_0 to $\gamma_0\gamma_5$.) The line is drawn with unit log-log slope to test proportionality to $\alpha_s^2 a^2$.

matrices can be diagonalized exactly giving four identical actions, each of them with one spin per site. If we keep only one of the actions, we reduce the lattice fermion degrees of freedom by a factor of four, which still leaves us with four fermion doublers. These residual degrees of freedom are called "tastes." Without further intervention, they would overcount the sea quark effects by a factor of four. To get approximately the correct counting, we replace the fermion determinant by its fourth root for each of the desired flavors:

$$Z_{\text{stagg}} = \int [dU] \, \exp[-S_G(U)] \prod_i \det[M_i(U)]^{1/4}. \quad (22)$$

In the continuum limit at nonzero quark masses, the eigenvalues of the determinant cluster in increasingly tighter quartets as expected from fermion doubling [18]. Then we have an $SU(4)$ taste symmetry, so taking the fourth root is equivalent to using only one of them as a sea quark species. This procedure has generated considerable controversy. Although there is no rigorous proof that the method is valid, all indications so far are that the approximation is under control as long as we take the continuum limit before we take the quark masses to zero or fit data to a chiral model with taste symmetry breaking properly included [19], in which case the limits are completely under control.

At nonzero lattice spacing the taste symmetry is broken, which introduces lattice artifacts. For example,

mesons composed of a valence quark and antiquark come in nondegenerate taste multiplets of sixteen tastes. In the continuum limit they are degenerate.

The asqtad [20–28] and p4fat3 [29,30] improvements of the staggered fermion formalism eliminate errors of $\mathcal{O}(a^2)$ in the quark dispersion relation and suppress taste splitting significantly. The asqtad suppression is somewhat better, presumably because it eliminates all tree level $\mathcal{O}(a^2)$ errors. The recently proposed HISQ action does still better [31]. In fig. 2 we compare the predicted and measured scaling of the splitting in the asqtad pion taste multiplet.

Taste splitting can also be reduced simply by replacing the gauge-link matrices in the action by smoothed gauge links —for example the Dublin "stout links" [32]. Unlike the asqtad approach, this method does not eliminate $\mathcal{O}(a^2)$ errors systematically. Thus the free quark dispersion relation is still unimproved.

Is taste symmetry breaking really a problem for thermodynamics? It is believed to be most dramatic for the pion and less noticeable for more massive states [33]. One could argue that close to the crossover temperature and away from the critical point, so many excited states participate, as in the hadron resonance gas model, that pions do not matter much. But if we approach the critical point at fixed lattice spacing, taste splitting is likely to have a strong effect on the critical behavior: we may even get a chiral-symmetry restoring transition in the wrong universality class. And certainly at quite low temperatures where pions dominate the statistical ensemble, taste splitting makes a difference.

Taste symmetry breaking also complicates the definition of the "physical" quark mass in a thermodynamics simulation. At zero temperature it is traditional to adjust the up and down quark masses so that the Goldstone pion (the lightest one in the taste multiplet) has the physical pion mass. This is legitimate, because we may restrict our attention to Green's functions whose external legs are the Goldstone pion. In a thermodynamics simulation, however, all members of the taste multiplet participate in the thermal ensemble. Thus it is more appropriate to tune the average multiplet mass, e.g., the rms pion mass to the physical pion mass. At a nonzero lattice spacing, the multiplet splitting may be so large, that goal is unreachable. In that case the physical point is reached only by reducing the lattice spacing together with the light quark mass. It is simply incorrect to claim a thermodynamics calculation is done at a physical pion mass when the rms mass is still much higher.

2.4.3 Domain-wall fermions

Neither the Wilson fermion formulation, including the clover-improved and twisted-mass version, nor the staggered fermion formulation are entirely satisfactory discretizations of fermions. Wilson fermions explicitly break chiral symmetry and its recovery requires a fine tuning. Staggered fermions, while preserving a remnant of chiral symmetry, have a remaining doubling problem, requiring the fourth-root trick, which is still somewhat controversial.

A more sophisticated, somewhat indirect and more costly discretization of fermions goes under the name of "domain-wall fermions" and was developed by Kaplan [34] and by Furman and Shamir [35]. Furman and Shamir's construction has become standard. An additional, fifth dimension of length L_s is introduced and one considers 5d Wilson fermions with no gauge links in the fifth direction, and the 4d gauge links independent of the fifth coordinate, s,

$$
S_{DW} = \sum_{s=0}^{L_s-1} \sum_x \bar{\psi}(x,s) \left\{ \sum_\mu \left(\gamma_\mu \nabla_\mu - \frac{1}{2} \Delta_\mu \right) \psi(x,s) \right.
$$
$$
\left. - M\psi(x,s) - P_-\psi(x,s+1) - P_+\psi(x,s-1) \right\}, \quad (23)
$$

where $P_\pm = \frac{1}{2}(1\pm\gamma_5)$ are chiral projectors and we have set $r = a = 1$. The parameter M, often referred to as domain-wall height, is introduced here with a sign opposite that of the usual mass term for Wilson fermions (eq. (19)). It needs to be chosen in the interval $0 < M < 2$. For free fermions the optimal choice is $M = 1$, while in the interacting case M should be somewhat larger. The fermion fields satisfy the boundary condition in the fifth direction,

$$
P_-\psi(x,L_s) = -m_f P_-\psi(x,0),
$$
$$
P_+\psi(x,-1) = -m_f P_+\psi(x,L_s-1), \quad (24)
$$

where m_f is a bare quark mass.

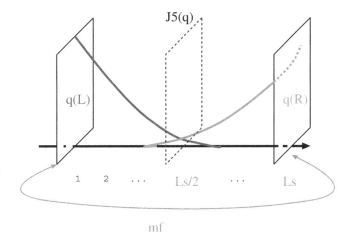

Fig. 3. Sketch, courtesy of Taku Izubuchi, of the domain-wall fermion setup. Left- and right-handed modes are exponentially bound to the left and right domain walls. The residual mass m_{res} is determined from an axial Ward identity applied in the center slice.

The domain-wall fermion action, eq. (23), has 5d chiral zero modes Ψ bound exponentially to the boundaries at $s = 0$ and $s = L_s - 1$, which are identified with the chiral modes of 4d fermions as

$$
q^R(x) = P_+\psi(x, L_s - 1), \qquad q^L(x) = P_-\psi(x, 0),
$$
$$
\bar{q}^R(x) = \bar{\psi}(x, L_s - 1)P_-, \qquad \bar{q}^L(x) = \bar{\psi}(x, 0)P_+. \quad (25)
$$

The left- and right-handed modes q^L and q^R do not interact for $m_f = 0$ when $L_s \to \infty$ and the domain-wall action has a chiral symmetry. At finite L_s the chiral symmetry is slightly broken. A popular measure of the chiral symmetry breaking is called "residual mass", m_{res}. It is determined from the axial Ward identity applied at the midpoint between the two domain walls, as sketched in fig. 3. This residual mass was expected to fall off exponentially in L_s. But, due to lattice artifacts of Wilson fermions with large negative mass, there is a contribution to m_{res} that decreases only like $1/L_s$ [36,37]. An example from a recent dynamical domain-wall fermion simulation [38] is shown in fig. 4. Nevertheless, often $L_s = \mathcal{O}(10\text{–}20)$ is large enough to keep the chiral symmetry breaking negligibly small, especially at smaller lattice spacing (weaker coupling).

Domain-wall fermions, therefore, solve, or at least substantially alleviate, explicit chiral symmetry breaking without a doubling problem. The price is a computational cost roughly a factor of L_s larger than that for Wilson-type fermions.

Early, $N_\tau = 4$ nonzero-temperature domain-wall fermion simulations suffered from large residual mass, since the lattice spacing in the transition/crossover region is large, leading to much heavier quarks than desired [39]. More recent simulations with $N_\tau = 8$, still using $L_s = 32$ and even 96 are described in [40]. Even for the $L_s = 32$ simulations with $N_t = 8$ the residual mass is uncomfortably large in the transition region, and getting worse at lower temperatures, corresponding to smaller β as shown

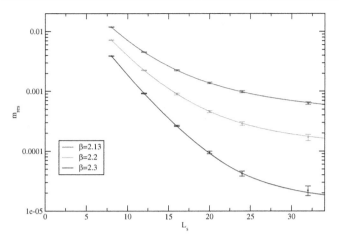

Fig. 4. Plot of the residual mass m_{res} as a function of L_s showing its desired suppression with increasing L_s and increasing inverse gauge coupling β. Also shown are fits to an exponential fall-off plus a $1/L_s$ contribution, from a recent $(2+1)$-flavor dynamical domain-wall fermion simulation [38].

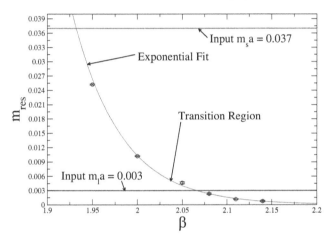

Fig. 5. Residual mass m_{res} for the recent nonzero-temperature simulations on $N_\tau = 8$ lattices with $L_s = 32$ [40]. At lower β, corresponding to lower temperatures, m_{res} increases rapidly, and is larger than the input light quark mass already in the transition region.

in fig. 5, since one would like m_{res} to be small compared with the input light quark masses.

2.4.4 Overlap fermions

Related to the domain-wall fermions of the previous subsection are the so-called overlap fermions developed by Narayanan and Neuberger [41,42]. They retain a complete chiral symmetry without the doubling problem, albeit again at substantial additional computational cost.

The overlap Dirac operator for massless fermions can be written as [42],

$$aD_{ov} = M\left[1 + \gamma_5\varepsilon\left(\gamma_5 D_W(-M)\right)\right], \quad (26)$$

where $D_W(-M)$ is the usual Wilson-Dirac operator with negative mass $m = -M$. As with domain-wall fermions

$0 < M < 2$ should be used. For a Hermitian matrix X, $\varepsilon(X)$ is the matrix sign function, that can be defined as

$$\varepsilon(X) = \frac{X}{\sqrt{X^2}}. \quad (27)$$

Using the fact that $\varepsilon^2(X) = 1$ it is easy to see that the Neuberger-Dirac operator satisfies the so-called Ginsparg-Wilson relation [43],

$$\{\gamma_5, D_{ov}\} = aD_{ov}\gamma_5 R D_{ov}, \quad (28)$$

with $R = 1/M$. Equivalently, when the inverse of D_{ov} is well defined, it satisfies

$$\{\gamma_5, D_{ov}^{-1}\} = a\gamma_5 R. \quad (29)$$

Chiral symmetry, in the continuum, implies that the massless fermion propagator anticommutes with γ_5. As seen above, the massless overlap propagator violates this only by a local term that vanishes in the continuum limit. According to Ginsparg and Wilson this is the mildest violation of the continuum chiral symmetry on the lattice possible. Lüscher [44] has shown that any Dirac operator satisfying the Ginsparg-Wilson (G-W) relation (28) has a modified chiral symmetry at finite lattice spacing,

$$\delta\psi = i\epsilon\gamma_5\left(1 - \frac{a}{2M}D\right)\psi, \quad \delta\bar{\psi} = i\epsilon\bar{\psi}\left(1 - \frac{a}{2M}D\right)\gamma_5. \quad (30)$$

or

$$\delta\psi = i\epsilon\gamma_5\left(1 - \frac{a}{M}D\right)\psi = i\epsilon\hat{\gamma}_5\psi, \quad \delta\bar{\psi} = i\epsilon\bar{\psi}\gamma_5, \quad (31)$$

with $\hat{\gamma}_5 = \gamma_5(1 - \frac{a}{M}D)$ satisfying $\hat{\gamma}_5^\dagger = \hat{\gamma}_5$ and, using the G-W relation, eq. (28), $\hat{\gamma}_5^2 = 1$.

So far only one exploratory study, on a $6^3 \times 4$ lattice, of nonzero-temperature overlap fermions has been done [45]. The main difficulty and computational cost for overlap fermions comes from the numerical implementation of the matrix sign function, eq. (27).

2.5 Cutoff effects

In selecting a fermion formalism for a thermodynamics study, it is important to be aware of possible lattice artifacts (cutoff effects). There are two important categories of artifacts. One comes from an imperfect rendering of chiral symmetry. The other, from the free quark dispersion relation.

It is obviously important to get the chiral symmetry right if we are simulating close to a chiral phase transition. Each action has its problems with chiral symmetry. For staggered fermions the taste splitting interferes. For Wilson fermions, the chiral symmetry is explicitly broken at nonzero lattice spacing. For these actions the obvious remedy is to reduce the lattice spacing. For domain-wall fermions, chiral symmetry is broken to the extent the fifth dimension is not infinite, and, for overlap fermions, chiral

Table 1. Continuum limit scaling behavior of free massless quarks in various lattice formulations, based on an expansion (eq. (32)) of the pressure in powers of $1/N_\tau^2$ from [46]. Shown are ratios of the expansion coefficients to the ideal, leading Stefan-Boltzmann coefficient. A small ratio indicates good scaling.

Action	A_2/A_0	A_4/A_0	A_6/A_0
Standard staggered	248/147	635/147	3796/189
Naik	0	−1143/980	−365/77
p4	0	−1143/980	73/2079
Standard Wilson	248/147	635/147	13351/8316
Hypercube	−0.242381	0.114366	−0.0436614
Overlap/ domain wall	248/147	635/147	3796/189

symmetry is broken to the extent the matrix sign function is only approximated in numerical simulations. For the latter two chiral actions, this type of error can be reduced without also reducing the lattice spacing.

At high temperatures where quarks are effectively deconfined, it would seem important to have a good quark dispersion relation, so, for example, we get an accurate value for the energy density and pressure. This artifact can be studied analytically for free fermions. Recently, Hegde *et al.* [46] looked at deviations from the expected free-fermion Stefan-Boltzmann relation for the pressure p as a function of $1/N_\tau^2$ (equivalently a^2) and chemical potential μ/T:

$$\frac{p}{T^4} = \sum_{k=0}^{\infty} A_{2k} P_{2k}(\mu/\pi T) \left(\frac{\pi}{N_\tau} \right)^{2k}, \quad (32)$$

where $P_{2k}(\mu/\pi T)$ is a polynomial normalized so that $P_{2k}(0) = 1$. The leading term A_0 is the Stefan-Boltzmann term. The ratios of higher coefficients A_{2k}/A_0 measure the strength of the cutoff effects. These terms determine the ability of the action to approximate the continuum free fermion dispersion relation, and they are useful in comparing actions to the extent free quarks are relevant in an interacting plasma. Table 1 reproduces their results for a variety of actions. We see that the hypercube action [47] has pleasingly small coefficients. The Naik (asqtad) and p4 (p4fat3) actions remove the second-order term as designed, but the p4 action is better at sixth order. The standard (unimproved) staggered action (regardless of gauge-link smearing) does as poorly as does the standard (and clover-improved) Wilson actions. The overlap and domain-wall actions constructed from the standard Wilson kernel unfortunately inherit its poor behavior. Improving the kernel fermion action would help to reduce these cutoff effects.

3 Determining the transition temperature

We want to know the temperature of the transition from confined hadronic matter to a quark-gluon plasma for two obvious reasons: to interpret experimental data and to understand QCD as a field theory. If the transition is only a crossover, a likely possibility for QCD at the physical value of the quark masses as discussed below, and a true phase transition occurs only at unphysical values of the quark masses, then these two purposes diverge. A crossover temperature is imprecise, so its meaning could vary with the observable, but one can at least speak of a range of temperatures over which phenomenologically interesting changes take place, or one could choose one observable to identify a temperature. A true phase transition has a precise temperature defined by the singularity of the partition function, and all observables capable of producing a signal should agree about the temperature.

In this section we discuss a variety of observables commonly used to detect the transition. In the following sections we discuss what we have learned from them about the phase structure of QCD.

Two observables are traditionally used to determine the temperature of the transition: the Polyakov loop and the chiral condensate. The Polyakov loop is a natural indicator of deconfinement. The chiral condensate is an indicator of chiral symmetry restoration.

3.1 Polyakov loop and the free energy of color screening

The Polyakov loop is an order parameter for a high-temperature, deconfining phase transition in QCD in the limit of infinite quark masses. At finite quark masses it is no longer an order parameter, but it is still used to locate the transition. It is built from the product of timelike gauge-link matrices. It is the expectation value of the color trace of that product:

$$L(\mathbf{x}, a, T) = \left\langle \text{Tr} \prod_{\tau=0}^{N_\tau - 1} U_0(\mathbf{x}, \tau) \right\rangle. \quad (33)$$

This quantity is gauge invariant because the combined boundary conditions for gluon and fermion fields require that gauge transformations be periodic under $t \to t + N_\tau$. Translational invariance insures that it is independent of $\mathbf{x}$. It can be shown that the Polyakov loop measures the change in the free energy of the ensemble under the introduction of a static quark (excluding its mass).

$$L(a, T) = \exp[-F_L(a, T)/T]. \quad (34)$$

In that sense the Polyakov loop is a useful phenomenological quantity as we now explain. When a static quark is introduced it must be screened so that the ensemble remains a color singlet. At low temperatures, screening is achieved by binding to it the lightest antiquark, forming a static-light meson. The free energy cost then consists of the self-energy of the static charge, the binding energy, and the self-energy of the light quark. In the quark plasma, color neutrality is achieved through a collective shift of the plasma charges, as in Debye screening in an ordinary electrical plasma. Aside from the self-energy of the static quark, which is the same at all temperatures, the additional free energy cost is small. So we expect $F_q(a, T)$ to decrease abruptly in the transition from the confining regime to the plasma regime.

The static-quark self-energy diverges as $1/a$ in the limit of small lattice spacing, so it is convenient to remove it from the definitions of the free energy and the Polyakov loop:

$$F_L(a, T) = F_{\text{static}}(a) + F_q(T),$$
$$L_{\text{renorm}}(T) = \exp[-F_q(T)/T]. \tag{35}$$

Figure 6 illustrates the free energy from a recent lattice simulation. (Here and elsewhere, the temperature scale is given in MeV and in units of the Sommer parameter [48], $r_0 \approx 0.467$ fm. The latter is defined in terms of the potential $V(r)$ between a heavy quark and antiquark. It is the distance where $r^2 dV(r)/dr = 1.65$.) The renormalized free energy behaves as expected.

If we take the masses of all the quarks to infinity, we arrive at the pure $SU(3)$ Yang-Mills ensemble, which has a first-order deconfining phase transition with zero $L(a, T)$ at low temperatures and nonzero at high temperatures. The free energy is correspondingly infinite for $T < T_c$ and finite above. In this limit the Polyakov loop is a true order parameter for the transition.

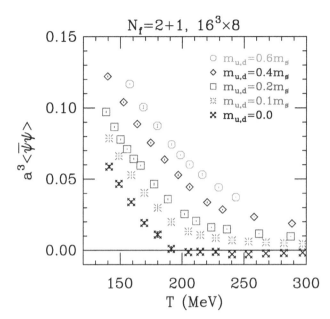

Fig. 7. Chiral condensate *vs.* temperature in MeV units (r_0 scale) for $N_\tau = 8$ from [52] using the asqtad fermion formulation. Measurements were taken along lines of constant physics with a range of light, degenerate up- and down-quark masses m_{ud} specified in the legend as a fraction of the strange quark mass m_s. An extrapolation to zero quark mass is also shown.

3.2 Chiral condensate

3.2.1 Chiral symmetry restoration

The second traditional observable is the chiral condensate. It is the order parameter for a high-temperature, chiral-symmetry restoring phase transition at zero up- and down-quark masses. At nonzero quark masses, it is no longer an order parameter, but, like the Polyakov loop, it is used as an indicator of the transition. It is defined for each quark flavor i as the derivative of the thermodynamic potential $\ln Z$ with respect to the quark mass,

$$\langle \bar{\psi}_i(x)\psi_i(x) \rangle = \frac{T}{V}\frac{\partial \ln Z}{\partial m_i} = \frac{T}{V}\left\langle \text{Tr}\, M_i^{-1} \right\rangle, \tag{36}$$

or the expectation value of the trace of the inverse of the fermion matrix. When the u- and d-quark masses both vanish, QCD has a $U(1) \times SU(2) \times SU(2)$ chiral symmetry, which is spontaneously broken at low temperatures to $U(1) \times SU(2)$, *i.e.*, the familiar baryon number and isospin symmetries. At high temperatures the full chiral symmetry is restored. The chiral condensate $\langle \bar{\psi}\psi \rangle$ is the order parameter of the broken symmetry. It is nonzero at low temperatures and zero at high temperatures. With only two flavors, the phase transition is expected to be second order, so the chiral condensate is continuous at the transition. When the quark masses are small, but nonzero, as they are in nature, the symmetry is explicitly broken and the chiral condensate does not vanish at high temperatures, but it is small.

Figure 7 illustrates the behavior of the chiral condensate from a recent lattice simulation with two light quark

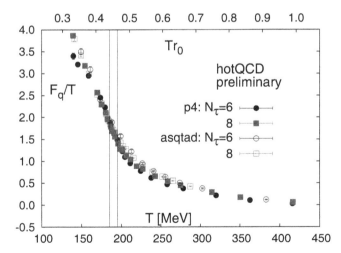

Fig. 6. Renormalized screening free energy of a static quark (from the renormalized Polyakov loop) *vs.* temperature in MeV units (bottom scale) and r_0 units (top scale) for $N_\tau = 6$ and 8 from a HotQCD study comparing p4fat3 and asqtad staggered fermion formulations [49–51]. Measurements are taken along a line of constant physics with $m_{ud} = 0.1 m_s$. The vertical bands here and in HotQCD figures below indicate a temperature range 185–195 MeV and serve to facilitate comparison.

flavors and one massive strange quark. Measurements were taken along "lines of constant physics", *i.e.*, curves in the space of the bare parameters (gauge coupling and quark masses) along which the pion and kaon masses are approximately constant, whether or not they have their correct experimental values. The extrapolation to zero light quark mass appears to be consistent with the expected behavior of this chiral order parameter.

3.2.2 Chiral multiplets

The restoration of chiral symmetry leads to symmetry multiplets in the hadronic spectrum. At low temperatures, where the symmetry is spontaneously broken, the spectrum consists of the familiar hadrons. At high temperatures, where the symmetry is restored, they may have analogs as resonant plasma excitations, at least not too far above the crossover temperature. They also control screening in the plasma in analogy with the Yukawa interaction. (See sect. 7.2.)

When the light quarks are massless, spontaneous symmetry breaking requires that the pion be massless. If the symmetry is restored at high temperatures, the pion, suitably defined as a state, acquires a mass. Of course, in nature, the light quarks are not massless, so the symmetry is only approximate, and the pion has a small mass at low temperatures.

Another consequence of restoring the chiral symmetry with massless u and d quarks is that all hadronic states involving those quarks would fall into larger symmetry multiplets. Thus, for example, the three pions become degenerate with the f_0, the three a_0's become degenerate with the η, and nucleons become degenerate with parity partner nucleons.

The classical QCD Lagrangian suggests a further $U(1)$ chiral symmetry, which would conserve a flavor-singlet axial charge. This symmetry is broken at the quantum level. This quantum phenomenon is called the Adler-Bell-Jackiw axial anomaly [53]. Whether the strength of the anomaly decreases in conjunction with the high-temperature transition is an open question.

If the anomaly also vanishes, the eight meson states listed above fall into a single degenerate supermultiplet. Again, if the light quarks are not precisely massless or the anomaly does not completely vanish, these statements are only approximate.

Whether or not hadron-like resonances are observable in experiments, the multiplets appear, nonetheless, in calculations, most notably in simulations of hadronic screening.

3.2.3 Singularities of the chiral condensate

Although we require a numerical simulation to determine the chiral condensate, from general considerations we can predict some of its singularities at small quark mass m

and small lattice spacing a:

$$\langle \bar{\psi}\psi(a, m, T) \rangle \sim$$
$$\begin{cases} c_{1/2}(a, T)\sqrt{m} + c_1 m/a^2 + \text{reg.}, & T < T_c, \\ c_1 m/a^2 + c_\delta m^{1/\delta} + \text{reg.}, & T = T_c, \\ c_1 m/a^2 + \text{reg.}, & T > T_c. \end{cases} \quad (37)$$

Knowing the behavior of the condensate, and in particular its singularities, is important for locating the phase transition. The m/a^2 singularity is easily derived in perturbation theory from a one-quark-loop diagram. The $\sqrt{m}$ singularity at low temperatures arises in chiral perturbation theory at one-loop order. In this case the pion makes the loop. It is an infrared singularity caused by the vanishing of the pion mass at zero quark mass [54]. Thus it appears only in the confined phase where the pion is massless. If we take $T \to 0$ before $m \to 0$ the square root singularity is replaced by the usual chiral $\log(m)$. The term $m^{1/\delta}$ is the expected critical behavior at the transition temperature. (For the expected 3d $O(4)$ universality class, $\delta = 0.56$.) The RBC-Bielefeld group discusses evidence for the expected mass dependence [54].

In a calculation with three quarks with masses $m_u = m_d = m_\ell$ and m_s, it is convenient for comparing results of different calculations to eliminate the ultraviolet divergence by taking a linear combination of the light quark and strange quark chiral condensates

$$D_{\ell,s}(T) = \langle \bar{\psi}\psi \rangle|_\ell - \frac{m_\ell}{m_s}\langle \psi\psi \rangle|_s,$$
$$\Delta_{\ell,s}(T) = D_{\ell,s}(T)/D_{\ell,s}(T=0). \quad (38)$$

The ratio $\Delta_{\ell,s}$ of the high-temperature and zero-temperature value also eliminates a common scalar-density renormalization factor Z_S. This is the quantity plotted in fig. 8 from a recent simulation. It shows the expected dramatic fall-off at the crossover.

3.3 Other observables

Susceptibilities are often used as indicators of a phase transition. They measure fluctuations in the related observables. Since a transition or crossover is usually accompanied by fluctuations in an order parameter, the related susceptibilities tend to peak there.

3.3.1 Quark number susceptibility

In the low-temperature phase, fluctuations in quark number are suppressed by confinement for the same reason that the free energy of screening of a static quark is large there. At high temperatures, fluctuations are common. There can also be cross correlations. The relevant observable for a quark of flavor i is the expectation $\langle N_i^2/V \rangle$ for spatial volume V and total quark number N_i. This is the quark number susceptibility. It controls event-by-event fluctuations in the associated flavor in heavy-ion collisions.

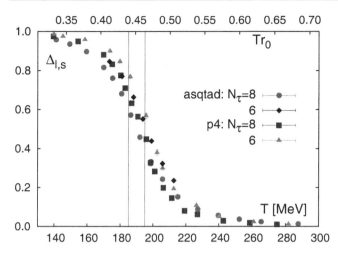

Fig. 8. To give an indication of its variation with lattice spacing, we plot the chiral condensate difference ratio *vs.* temperature in MeV units (bottom scale) and r_0 units (top scale) for $N_\tau = 6$ and 8 from a HotQCD study. Results are given for both the p4fat3 and asqtad staggered fermion formulations [51]. Measurements are taken along a line of constant physics with $m_{ud} = 0.1m_s$.

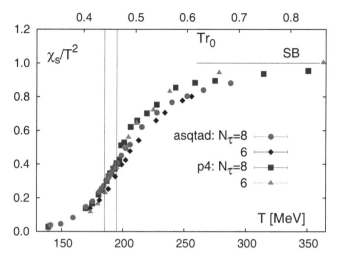

Fig. 9. Strange quark number susceptibility divided by the square of the temperature *vs.* temperature in MeV units (bottom scale) and r_0 units (top scale) for $N_\tau = 6$ and 8. Measurements are taken along a line of constant physics with $m_{ud} = 0.1m_s$. Results are from a HotQCD study comparing p4fat3 and asqtad staggered fermion formulations [51].

For flavors i and j the generalized susceptibility (including cross correlations) is

$$\chi_{ij} = \langle N_i N_j / V \rangle = \frac{T}{V} \frac{\partial^2 \ln Z}{\partial \mu_i \partial \mu_j}. \qquad (39)$$

We discuss the Taylor expansion of this observable in μ_i in sect. 6.5.

Figure 9 illustrates the behavior of the strange quark number susceptibility χ_{ss}. It shows an abrupt rise at the crossover. Because it has a relatively high signal to noise ratio, this quantity is often used to define the crossover temperature.

We can transform the generalized quark number susceptibility χ_{ij} from the flavor basis to the basis in which the isospin I, hypercharge Y, and baryon number B are diagonal. The resulting quantities are shown in fig. 10. The diagonal susceptibilities all show the expected abrupt rise at the crossover temperature. The offdiagonal susceptibility $\chi_{Y,B}$ shows a small nonzero value above the crossover. The positive correlation between hypercharge and baryon number at these temperatures can either be understood in terms of fluctuations in light quark degrees of freedom or in terms of persistent three-quark baryon states: light up and down quarks have positive baryon number (1/3) and hypercharge (1/3) and their antiquarks have the opposite values. In both cases their fluctuations lead to a positive correlation. Strange quarks have positive baryon number (1/3) but negative hypercharge (−2/3). They would lead to a negative correlation, but because of their higher mass, they are less prevalent. So we are left with a net positive correlation. At higher temperatures the mass difference is irrelevant and the correlations cancel. Similar arguments can be made for three-quark baryonic states, where non-strange baryons are more prevalent than strange baryons.

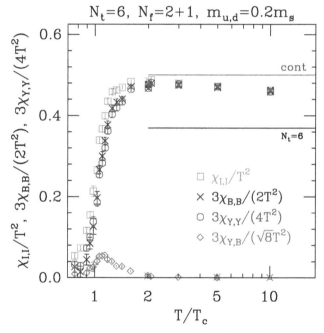

Fig. 10. Chiral susceptibility matrix in the I, Y, B basis. divided by the square of the temperature *vs.* temperature in units of the crossover temperature T_c for $N_\tau = 6$. Measurements are taken along a line of constant physics with $m_{ud} = 0.2m_s$ from [52].

In fig. 11 we show a computation of baryon (χ_q) and isospin (χ_I) quark number susceptibilities from a recent computation with two flavors of clover-improved Wilson fermions on $16^3 \times 4$ lattices [55], using a different normalization from that of fig. 10. The cutoff effects for this Wilson fermion simulation are seen to be significantly larger than with improved staggered fermions, as expected from table 1.

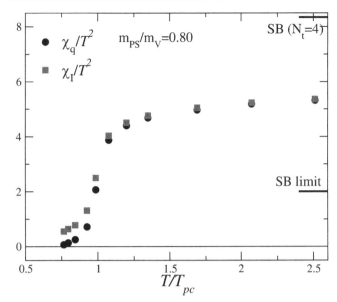

Fig. 11. Quark number susceptibilities with Wilson fermions on $N_\tau = 4$ lattices along a line of constant physics with pseudoscalar to vector meson mass ratio $m_{PS}/m_V = 0.8$ from [55].

3.3.2 Chiral susceptibility

The various chiral susceptibilities are based on the second derivative of the thermodynamic potential with respect to the quark masses

$$\chi_{ij} = \frac{T}{V} \frac{\partial^2 \ln Z}{\partial m_i \partial m_j}. \tag{40}$$

For two equal-mass light quarks u and d, the derivatives can be converted to expectation values of products of the inverse of the fermion Dirac matrix M for those species. The commonly reported susceptibilities are the "disconnected" chiral susceptibility

$$\chi_{\text{disc}} = \frac{T}{V} \left[\langle (\text{Tr}\, M^{-1})^2 \rangle - \langle \text{Tr}\, M^{-1} \rangle^2 \right], \tag{41}$$

the connected chiral susceptibility

$$\chi_{\text{conn}} = \frac{T}{V} \langle \text{Tr}\, M^{-2} \rangle, \tag{42}$$

the isosinglet susceptibility

$$\chi_{\text{sing}} = \chi_{\text{conn}} + 2\chi_{\text{disc}}, \tag{43}$$

and the isotriplet susceptibility

$$\chi_{\text{trip}} = 2\chi_{\text{conn}}. \tag{44}$$

Figure 12 gives an example of the peak in the disconnected chiral susceptibility at the crossover.

Since the chiral susceptibility is the derivative of the chiral condensate with respect to quark mass, one can immediately derive its singularities from the expressions

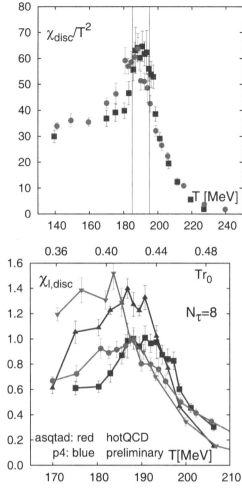

Fig. 12. Upper panel: disconnected light quark susceptibility *vs.* temperature in MeV (r_0) units (bottom scale). Lower panel: closeup of the peak region. Lines merely connect the points. Red circles and downward-pointing triangles, asqtad fermions. Blue squares and upward-pointing triangles, p4fat3. Squares and circles are along a line of constant physics with $m_{ud} = 0.1 m_s$, and triangles, with $m_{ud} = 0.05 m_s$. All results are HotQCD preliminary [49,56,50].

for the condensate in eq. (37):

$$\chi_{\text{sing}}(a, m, T) \sim$$
$$\begin{cases} c_{1/2}(a,T)/(2\sqrt{m}) + c_1/a^2 + \text{reg.}, & T < T_c, \\ c_1/a^2 + (c_\delta/\delta)m^{1/\delta - 1} + \text{reg.}, & T = T_c, \\ c_1/a^2 + \text{reg.}, & T > T_c. \end{cases} \tag{45}$$

The RBC-Bielefeld group discusses numerical evidence for the expected mass dependence [54]. Trends in fig. 12 are consistent with these expectations. In the limit of zero quark mass, this quantity is infinite below the transition and finite above. In the continuum limit it has a temperature-independent ultraviolet divergence. Thus the Budapest/Wuppertal group proposes subtracting the zero-temperature value, multiplying by the square of the bare quark mass, and dividing by the fourth power of the temperature [32]:

$$m^2 \Delta \chi_{\text{disc}}(a, m, T)/T^4. \tag{46}$$

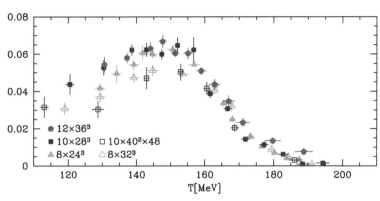
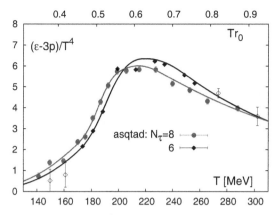

Fig. 13. Left panel: renormalized chiral susceptibility *vs.* temperature (f_K scale) from [57]. Right panel: interaction measure *vs.* temperature (r_0 scale) from [51]. (See the definition of this quantity in sect. 6.2.) Note that the interaction measure peaks at about 20 MeV above the crossover.

where $\Delta\chi_{\text{disc}}(a,m,T) = \chi(a,m,T) - \chi(a,m,0)$. The m^2 cancels the scalar-density renormalization factor. Of course, this quantity vanishes in the zero-mass limit.

3.4 Setting the temperature scale

In order to quote dimensionful lattice results in physical units, it is necessary to determine the lattice spacing in physical units. The calibration must be based on a quantity that is reliably determined in zero-temperature lattice simulations. Recent favorites are the splitting of Υ levels, the mass of the Ω^- baryon, and the light meson decay constants, such as f_π or f_K. These scale determinations are not guaranteed to agree at nonzero lattice spacing and at unphysical values of the quark masses. Indeed, there can be substantial differences. For example, for the asqtad action with a nearly physical strange quark mass, a light quark mass one tenth as heavy, and a lattice spacing of approximately 0.12 fm, the f_K scale gives a 15% lower temperature than the Υ splitting scale. For the same quark masses, at approximately 0.09 fm the discrepancy has decreased to 8%, consistent with an approximately $\mathcal{O}(a^2)$ scaling error. Of course, for any quantity of interest, thermodynamic or not, if possible, we would like to choose a scale according to which that quantity has only a small variation as the lattice spacing approaches zero.

Recent results from Aoki *et al.* [57] give a rather different temperature T_c for the crossover than the HotQCD Collaboration [51]. Aoki *et al.* locate the peak in their renormalized chiral susceptibility at around 150 MeV (f_K) for $N_\tau = 8$, 10, and 12. The HotQCD Collaboration puts the crossover closer to 190 MeV (r_0) for $N_\tau = 8$ and $m_{ud}/m_s = 0.1$. Here are possible reasons for the discrepancy:

– Much of the difference comes from the different choice of scale. The Budapest-Wuppertal Collaboration uses f_K to set the scale, and the HotQCD Collaboration uses the Sommer parameter r_0, calibrated ultimately from Υ splittings [58]. The scale discrepancy alone could explain about 30 MeV of the difference.

– Some of the discrepancy also comes from differences in lattice parameters. The Budapest-Wuppertal Collaboration uses a smaller lattice spacing and lighter light quark mass. The HotQCD Collaboration estimates an approximately 10 MeV (r_0 scale) downward shift in curves related to the equation of state in the continuum limit with physical quark masses. Some of that shift is visible in the right panel of fig. 13.

– Some may also come from differences in the fermion formulations. The Budapest-Wuppertal group use standard staggered fermions with stout gauge links. This approach reduces effects of taste splitting, but does not improve the quark dispersion relation as do the actions used by the HotQCD Collaboration. We do not know whether such differences would result in a shift in a peak position, however.

Whatever the differences, no matter how one sets the scale, one expects all methods to give the same results for the same observable in the continuum limit at physical quark masses. So for now we are left guessing the result of taking that limit. Since most of the present difference apparently comes from a choice of scale, it would help our guessing to know which scale is more suitable for thermodynamic quantities. We have seen that the chiral susceptibility suffers from peculiar singularities that may make it less suitable for locating the crossover temperature. Still, the left panel of fig. 13 suggests that it scales reasonably well in f_K units. For the phenomenology of heavy-ion collisions, quantities related more directly to deconfinement, such as the interaction measure (equation of state) and quark number susceptibility are important. As we can see from the right panel of fig. 13 the interaction measure seems to show better (but still imperfect) scaling in the r_0 scale. (Preliminary HotQCD results for the chiral susceptibility are shown in the r_0 scale in fig. 12.)

4 QCD phase diagram at zero density

4.1 General outline of the phase diagram

At infinite quark mass QCD becomes a pure Yang-Mills theory, which has a well-studied, weak, first-order deconfining phase transition [59]. As the quark masses are decreased, the first-order transition weakens further and devolves into a crossover, as indicated in fig. 14, which summarizes in *qualitative* terms the generally accepted phase structure at zero chemical potential in the flavors u, d, and s.

Close to zero quark mass, chiral perturbation theory applies, and quite general arguments can be made about the qualitative nature of the phase transition [60], depending on the number of quark flavors with zero mass and depending on what happens to the anomaly at the transition. With a nonzero anomaly and only two quark flavors the transition certainly occurs at zero u- and d-quark masses, and it is in the 3d $O(4)$ universality class, because of the $O(4)$ two-flavor chiral symmetry. If the strange quark is also massless, the chiral transition is first order, and, since first-order transitions are not usually removed by small symmetry-breaking perturbations, it persists as the quark masses are increased. Eventually, at sufficiently large u-, d-, and s-quark masses the system is too far from chiral and the first-order transition gives way to a second-order phase transition in the Ising or $Z(2)$ universality class: Ising, since at nonzero quark masses, there is no remaining chiral symmetry. In the $m_u = m_d$ vs. m_s plane a curve of such second-order transitions separates the first-order regime from the crossover regime as sketched in the upper panel of fig. 14.

The quantitative determination of the phase boundaries requires numerical simulation. What has emerged is that the second-order critical line occurs at quite small quark masses, where simulations are particularly challenging and especially sensitive to cutoff effects [61,62]. The lower panel of fig. 14 shows recent results from de Forcrand and Philipsen based on a calculation using unimproved staggered fermions with $N_\tau = 4$.

4.2 Order of the phase transition for physical quark masses

A key phenomenological question is whether there is a first-order phase transition at the physical value of the u-, d-, and s-quark masses or there is merely a crossover. All present evidence points to a crossover at zero chemical potential for these species. A recent, thorough investigation has been carried out by the Budapest-Wuppertal group [63]. They examine the conventional signal of the peak height in the chiral susceptibility, which they renormalize using eq. (46). If there is no phase transition (*i.e.*, only a crossover), the peak height should be asymptotically constant in the thermodynamic limit of an infinite lattice volume. If there is a first-order phase transition, the height is infinite, but it is limited in a finite volume by finite-size effects. Asymptotically, it scales linearly with the lattice volume L^3. If the transition is second order, the volume dependence is weaker, but the result is still infinite. The Budapest-Wuppertal group ran a simulation with conventional staggered fermions on stout links at $N_\tau = 4$, 6, 8, and 10. They analyzed their data in two steps. First, they extrapolated the inverse peak height to zero lattice spacing at fixed lattice aspect ratio LT, as

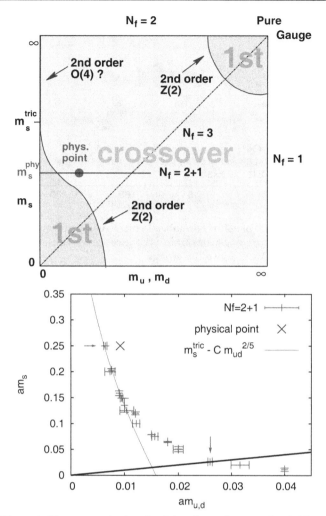

Fig. 14. Upper panel: sketch of the phase diagram for QCD at zero baryon density in $(2 + 1)$-flavor QCD as a function of the light quark masses showing regions where a high-temperature phase transition or crossover is expected. For a second-order phase transition, the universality class is shown. The physical point is plotted as a dot in the crossover region. Whether the expected tricritical strange quark mass m_s^{tric} is higher or lower than the physical strange quark mass m_s^{phys} is not yet firmly established. (Similar versions of this figure have appeared in the literature, including [64].) Lower panel: result of an actual measurement of a portion of the 2nd-order $Z(2)$ phase boundary at $N_\tau = 4$ from ref. [65]. The axes give bare quark masses in lattice units and the blue cross marks the physical point.

shown in the upper panels of fig. 15. Then they extrapolated the continuum values to infinite aspect ratio (thermodynamic limit). The result is compared in the lower panel of fig. 15 with predictions for a first-order phase transition and a phase transition in the 3d $O(4)$ universality class. The disagreement is a strong indication that there is no phase transition.

4.3 Order of the phase transition for two massless flavors

There is a related question of significant theoretical interest. When all quarks but the u and d are infinitely mas-

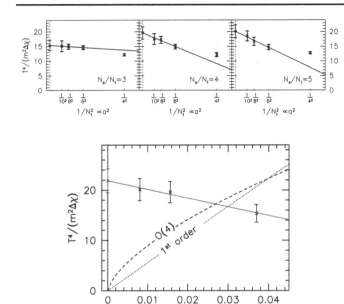

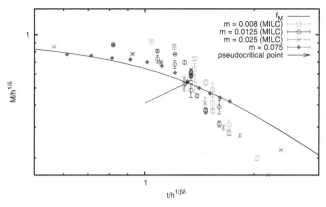

Fig. 16. A double lograrithmic plot showing strong deviations from $O(4)$ scaling in this parameter range for two flavors of staggered fermions using $N_\tau = 4$ lattices, collected in [66]. The function M is the chiral condensate (magnetization in the spin system), h the quark mass (external magnetic field) and β and δ are critical exponents. $t = 6/g^2 - 6/g_c^2$ plays the role of the reduced temperature.

Fig. 15. Results from [63]. Upper panels: inverse of the peak height in the renormalized disconnected chiral susceptibility *vs.* squared lattice spacing showing the extrapolation to zero lattice spacing. The lattice aspect ratio is varied from left to right. Lower panel: inverse of the peak height in the renormalized disconnected chiral susceptibility *vs.* inverse aspect ratio cubed showing the extrapolation to the thermodynamic limit. Also shown are predictions for a first-order phase transition and a second-order transition in the 3d $O(4)$ universality class.

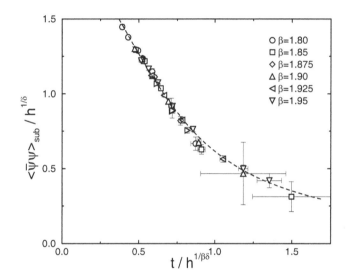

Fig. 17. $O(4)$ scaling, with linear scale, for two flavors of Wilson fermions using $N_\tau = 4$ lattices, from [16].

sive, we have a two-flavor theory, and, as we have observed above, as long as the chiral anomaly is not involved, we expect a critical point only at zero quark mass. Furthermore, since the two-flavor chiral symmetry $SU(2) \times SU(2) \simeq O(4)$, we expect the high-temperature deconfining critical point to be in the 3d $O(4)$ universality class.

This question has been investigated by several groups with somewhat contradictory results. Simulations with standard staggered quarks using $N_\tau = 4$ lattices, with large lattice spacing in the transition region, and hence potentially large lattice artifacts, as collected by T. Mendez [66], show some deviations from $O(4)$ scaling as shown in fig. 16. For $O(4)$ scaling, all data points should collapse to the curve in the figure. Two-flavor clover-improved Wilson fermion simulations [16], on the other hand, indicate good $O(4)$ scaling as seen in fig. 17.

Since staggered fermions, at the large lattice spacings in the transition region on high-temperature lattices with small N_τ, have quite large taste symmetry breaking, one might expect the transition to be in the $U(1) \times U(1) \simeq O(2)$ universality class, rather than the $O(4)$ one. More importantly, Kogut and Sinclair [67] argue that finite volume effects on the fairly small (spatial) lattices used are quite large. Indeed they found good agreement with $O(2)$ scaling, when taking the finite-volume effects into account as illustrated in fig. 18.

In contradiction with the theoretical expectations and the above-summarized numerical findings, D'Elia, Di Gi-

acomo, and Pica found indications of a first-order transition using an unimproved staggered fermion action and $N_\tau = 4$ [68]. It is important to check this conclusion with a more refined action. One should conclude that at present the order of the high-temperature transition with two massless flavors is still an open question.

4.4 The phase transition with a physical strange quark

Suppose, instead, we hold the strange quark mass at its physical value and then decrease the u- and d-quark masses toward zero. According to the qualitative picture in the upper panel of fig. 14, depending on where the tricritical point lies, we could 1) encounter a critical point and enter a first-order regime, or 2) we may have to go to zero quark mass to find a genuine phase transition. In the

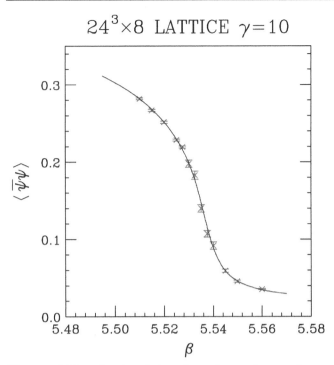

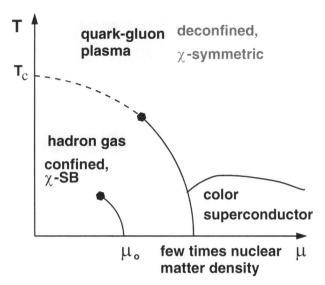

Fig. 18. $O(2)$ scaling in a finite volume for two flavors of massless staggered fermions with an irrelevant four-fermion interaction, from [67]. The curve comes from an $O(2)$ spin model simulation with "matched volume".

Fig. 19. Sketch of the expected phase diagram for $(2 + 1)$-flavor hadronic matter as a function of temperature and chemical potential at physical quark masses from [70]. The confined, chiral symmetry-broken phase lies in the lower left, separated from the deconfined, chirally symmetric phase by a pseudocritical crossover line (dashed) and first-order (solid) line of phase transitions. A critical point is indicated by a black hexagon. A nuclear matter phase transition occurs along a line extending from $\mu = \mu_0$. At higher densities a color superconducting phase is proposed.

lower panel we reproduce a result from de Forcrand and Philipsen suggesting the first alternative, but their results were obtained with an unimproved action at $N_\tau = 4$ for which we expect large cutoff effects.

As we have mentioned, cutoff effects complicate the determination of the phase boundary at small quark mass. This is especially likely to be true for simulations based on unimproved staggered fermions (even improved staggered fermions are not entirely immune), since for them it is important to take the continuum limit before taking the small quark mass limit. Otherwise, one risks being misled by lattice artifacts.

5 QCD Phase diagram at nonzero densities

5.1 Phenomenology

As the baryon number density is increased (*i.e.*, all the flavor chemical potentials are increased from zero), according to traditional arguments, there is a chiral-symmetry restoring phase transition along a line in the (μ, T)-plane when the u- and d-quark masses are zero, as sketched in fig. 19 [69]. This tradition is founded on two notions. The first argues that asymptotic freedom and consequently deconfinement should reign at very high temperatures and high chemical potential. The second argues that spontaneous chiral symmetry breaking occurs at zero chemical potential because, when fermions acquire a dynamical mass through symmetry breaking, the negative energy levels of the Dirac sea are lowered, lowering the vacuum en-

ergy. With a nonzero chemical potential the filled positive energy levels rise in energy, counteracting the advantage of a dynamical chiral mass, and consequently inhibiting spontaneous symmetry breaking [71].

At zero u- and d-quark mass chiral symmetry is exact. If chiral symmetry is restored above a critical chemical potential and it is spontaneously broken below, analyticity requires a phase transition. There are no such guarantees, however, when quark masses are not zero. Since we know from numerical simulation that at physical quark masses there is only a crossover at zero density, the critical line separating the chirally broken from the chirally restored phase must move away from the temperature axis as the quark masses are increased. It then terminates in a critical endpoint (T_E, μ_E). A crossover line then fills the gap from there to the temperature axis, as indicated by the dashed line in fig. 19. A key phenomenological question is whether the critical endpoint is experimentally accessible.

At still higher densities exotic phases have been proposed, including diquark condensates and color-flavor locked and superconducting phases [72,73]. These phases are, thus far, completely beyond the reach of current lattice simulations.

5.2 Lattice methods for nonzero densities

To confirm or refute these traditional arguments requires numerical simulation. Unfortunately, simulations at nonzero chemical potential are very difficult, since standard lattice methodology requires that the Feynman path integrand be treated as a positive probability measure. In $SU(3)$ gauge theory, the integrand becomes complex at a nonzero (real) chemical potential. This creates a fermion "sign problem" analogous to the fermion sign problem in condensed matter physics in strongly coupled electron systems away from half-filling. A solution to either problem would be beneficial to the other.

To see why the problem arises, consider the naive fermion Dirac matrix $M(U) = \gamma_\nu \nabla_\nu + m$. The lattice version of the gauge-covariant derivative ∇_ν is given by eq. (17). The terms in ∇_ν in the action allow the quark to hop to next neighbor sites in the positive and negative ν direction. Normally, hopping in all directions must have equal weight to preserve the discrete lattice symmetries of axis interchange, parity, time reversal, and charge conjugation. The fermion determinant is then real because taking its complex conjugate corresponds to reversing the direction of hopping, which has the same weight. But a positive nonzero chemical potential promotes quark hopping in the positive (imaginary) time direction and suppresses it in the negative time direction. This is naturally implemented by changing the covariant time derivative as follows:

$$\nabla_0 \psi(x) \rightarrow \frac{1}{2a} [U_0(x) e^{a\mu} \psi(x + \hat{0}a)$$
$$- e^{-a\mu} U_0^\dagger(x - \hat{0}a) \psi(x - \hat{0}a)]. \quad (47)$$

If a quark hops along a worldline that wraps completely around the lattice in the imaginary time direction, it accumulates N_τ factors of $\exp(a\mu)$, and the partition function receives a net enhancement $\exp(a\mu N_\tau) = \exp(\mu/T)$, the appropriate statistical weight for the addition of one quark to the grand canonical ensemble. A quark hopping backwards is interpreted as an antiquark, and its contribution is correspondingly suppressed, as it should be. With this imbalance the determinant is no longer guaranteed to be real. Instead it acquires a complex phase $\phi \propto \mu V$, i.e., roughly proportional to the lattice volume and the chemical potential.

A complex determinant creates additional problems for staggered fermions. With $2 + 1$ flavors of staggered fermions at nonzero densities, one requires the square root and fourth root of the fermion determinants. When the determinant is real, there is no phase ambiguity in the root. But when the determinant is complex, one has to choose the correct Riemann sheet. The ambiguities and an expensive remedy are discussed in [74]. To be safe, one is limited to small μ and volumes.

Over the years a number of methods have been proposed for treating a complex determinant. We give a brief account of the attempts. For recent reviews, see [75, 76].

5.2.1 Reweighting the fermion determinant

As a standard lattice Monte Carlo method, reweighting involves sampling the Feynman path integral according to one measure and then making adjustments to achieve

the effect of simulating with a slightly different measure [77, 78].

Let us see how this idea is applied to a simulation at nonzero chemical potentials μ_i, one for each flavor i. (To be precise, we are speaking of a quark number chemical potential. The baryon number potential is three times as large ($\mu_{Bi} = 3\mu_i$). The expectation value of an operator $\mathcal{O}$ is given by

$$\langle \mathcal{O} \rangle_\mu = \int [dU] \mathcal{O}(U) \exp[-S_G(U)] \prod_i \det[M_i(U, \mu_i)] / Z(\mu),$$
$$(48)$$

where $\mu = (\mu_1, \mu_2, \dots)$ and

$$Z(\mu) = \int [dU] \exp[-S_G(U)] \prod_i \det[M_i(U, \mu_i)]. \quad (49)$$

Since we cannot do importance sampling with the unsuitably complex determinant $\det[M(U, \mu)]$ in the measure, we can try to do it with the real determinant $\det[M(U, \mu = 0)]$. That is, we write

$$\langle \mathcal{O} \rangle_\mu = \langle \mathcal{O} R(U, \mu) \rangle_0 / \langle R(U, \mu) \rangle_0, \quad (50)$$

where $R(U, \mu)$ is the ratio of determinants that reweights the contributions to the integrand to compensate for the incorrect sampling measure:

$$R(U, \mu) = \det[M(U, \mu)] / \det[M(U, 0)]. \quad (51)$$

Similarly, we can reweight to imitate a change in any of the parameters of the action including the quark masses and gauge coupling. The reweighting factor R is simply the ratio of the intended and actual measures.

This procedure, often called the Glasgow method, is mathematically correct but numerically unstable. As the chemical potential moves away from zero, one is no longer doing importance sampling. In complex analysis this approach is similar to attempting to estimate a contour integral in the stationary phase approximation without going through the saddle point. The variance in the sampled values of the numerator and denominator in eq. (50) grows exponentially as the lattice volume increases, i.e., in the thermodynamic limit. The inevitable breakdown is forestalled by keeping the shift in parameters small, so by working at small μ.

A variant of this method uses the absolute value of the determinant for the sample weighting. The reweighting factor is then the phase [79]. This method has been applied only to small lattice volumes.

Fodor and Katz propose reweighting simultaneously in the gauge coupling g^2 and μ [80]. They argue that one achieves a better overlap with this method. For example, one might expect that if one moves along the crossover line in the (μ, T)-plane, the important integration domain might not change as rapidly as it would if one moves in some other direction. To stay on this line requires changing the gauge coupling along with the chemical potential. To locate the critical line, they follow Lee-Yang zeros of the partition function. (These zeros lie in the complex temperature or complex gauge-coupling plane. If there is a genuine phase transition, as the lattice volume is increased, they impinge on the real temperature axis and give rise to a singularity. If there is only a crossover, they stay harmlessly away from the real axis.) From this method they estimate the critical endpoint at $T = 160(3.5)\,\text{MeV}$ and $\mu_B = 3\mu = 360(40)\,\text{MeV}$ at physical quark masses using conventional staggered fermions [81]. This critical chemical potential is nearly a factor of two smaller than an earlier estimate at higher quark masses and smaller volumes [82]. Such sensitivity to the simulation parameters warrants further study.

5.2.2 Approximating the determinant with phase quenching

With degenerate up and down quarks, simulating with the "phase-quenched" or absolute value of the determinant and ignoring the phase completely is equivalent to giving the up-quark a positive chemical potential and the down-quark a negative chemical potential, so it is equivalent to simulating with an isospin chemical potential [83]. This procedure is numerically tractable, but to draw conclusions regarding the phase diagram with the standard chemical potential requires some justification. Kogut and Sinclair present the case in [84]. See also [85].

5.2.3 Simulating in the canonical ensemble

Another approach is to simulate in the canonical ensemble of fixed quark (baryon) number [86–89]. For simplicity, consider a single quark species. The canonical ensemble with quark number q is then obtained from the Fourier transform

$$Z_q = \int_0^{2\pi} \mathrm{d}\phi\, e^{-iq\phi} \int [\mathrm{d}U] \exp[-S_G(U)] \det[M(U,\mu)]\big|_{\mu/T=i\phi}.$$
(52)

The sign problem arises in the Fourier transform. As the quark number is increased for a given lattice volume and configuration, the Fourier component decreases rapidly and the sensitivity to oscillations worsens, so that any discrete approximation to the Fourier transform develops a severely large variance.

Meng *et al.* have recently proposed a new "winding number expansion" method that starts from the Fourier transform of the logarithm of the determinant, $\log(\det[M(U,\mu)]) = \mathrm{Tr}\log[M(U,\mu)]$ and proceeds via a Taylor expansion to generate the canonical partition function [90,91]. The method converges much better, but so far results are reported only for fairly large quark masses.

5.2.4 Simulating with an imaginary chemical potential

If we make the chemical potential purely imaginary, the fermion determinant becomes real, and a direct simulation [92] is possible. To recover results at a physical, real

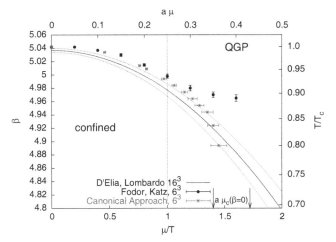

Fig. 20. From [88]. Critical line as a function of quark chemical potential and temperature for four degenerate flavors of unimproved staggered fermions at $N_\tau = 4$, bare quark mass $am = 0.05$, and for the spatial lattice volumes shown. Results from three methods are compared: the imaginary chemical potential approach of [93], the canonical ensemble approach of [88], and the multiparameter reweighting approach of [80]. A range of strong coupling values of the critical chemical potential $\mu_c(\beta = 0)$ is also indicated.

chemical potential, we must do an analytic continuation. The success of such a continuation depends on knowing the analytic form of the observable as a function of chemical potential. We do if the chemical potential is small enough that a Taylor expansion is plausible. So in the end, the imaginary potential method provides essentially the same information as an explicit Taylor expansion about zero chemical potential. Figure 20 from de Forcrand and Kratochvila [88] compares three methods for determining the critical line. Each result shown is based on the same unimproved $N_f = 4$ staggered fermion action. The methods agree reasonably well for $\mu/T < 1$. Note that this is a four-flavor simulation with a first-order phase transition, unlike the $(2 + 1)$-flavor case of fig. 19.

5.2.5 Taylor expansion method

For small chemical potential, we may carry out a Taylor expansion of the required observables in terms of the flavor chemical potentials at zero chemical potential [94,95]. Since all Taylor coefficients are evaluated at zero chemical potential, determining them is straightforward. However, the observables that give the coefficients are nontrivial. They involve products of various traces of the inverse fermion matrix. The traces are usually evaluated using stochastic methods. Furthermore, as the order of the expansion grows, the number of required terms grows factorially. Thus it is rare to find calculations as high as eighth order [96,97].

5.2.6 Probability distribution function method

The "probability distribution function" or "density-of-states" method is new and promising [98,99]. It is related to the reweighting method. A recent variant by Ejiri combines reweighting with a Taylor expansion. To explain the method we start with a simple case, defining the "density of states" or "probability distribution function" of the plaquette P with the Wilson gauge action and arbitrary fermion action:

$$w(P') = \int [dU]\, \delta(P' - P(U))\, \det[M(U,0)]\exp[-S_G(U)],$$
(53)

where $\delta(P'-P)$ is the Dirac delta function. It is defined like the partition function, but at a fixed value of the plaquette. The expectation value for an observable $\mathcal{O}(P)$ that depends only on P is then

$$\langle \mathcal{O}\rangle = \int dP'\, w(P')\mathcal{O}(P') \Big/ \int dP'\, w(P').$$
(54)

At nonzero μ we use reweighting to calculate the partition function:

$$Z(\mu) = \int dP\, R(P,\mu)w(P),$$
(55)

where the plaquette-restricted reweighting function $R(P,\mu)$ is

$$R(P,\mu) = \frac{\int [dU]\, \delta(P'-P(U))\, \det[M(U,\mu)]}{\int [dU]\, \delta(P'-P(U))\, \det[M(U,0)]},$$
(56)

i.e., the ratio at nonzero and zero μ. For the Wilson action, the gauge weight $\exp[-S_G(U)]$ depends only on P, so it cancels between numerator and denominator in $R(P,\mu)$. The distribution function $w(P)$ is still calculated at $\mu = 0$ according to (53).

The sign problem appears in the numeric evaluation of $R(P,\mu)$. Ejiri offers a way to overcome it [99]. His method begins with a generalization of the distribution function, making it depend on three variables: the plaquette P, the magnitude of the ratio of determinants $F(\mu) = \det M(\mu)/\det M(0)$, and the phase $\theta \equiv \operatorname{Im}\log\det M(\mu)$:

$$w(P',|F'|,\theta') = \int [dU]\, \delta(P'-P(U))\delta(|F'|-|F|)$$
$$\times \delta(\theta'-\theta)\det[M(U,0)]\exp[-S_G(U)],$$
(57)

Note that the real, positive weight factor in the integrand comes from the $\mu = 0$ action. For any value of μ the partition function is then

$$Z(\mu) = \int dP d|F| d\theta\, F(\mu)w(P,|F|,\theta),$$
(58)

where in place of the reweighting function R we now have simply $F(\mu)$ itself.

The next step relies on the key assumption that the distribution function $w(P',|F'|,\theta')$ is Gaussian in θ. Ejiri argues that this is plausible, at least for large volume. A further assumption for rooted staggered fermions is that the effect on the phase of taking the fourth root is simply to replace θ by $\theta/4$ in the Gaussian distribution. With these assumptions one can do the θ integration directly,

eliminating the sign problem. The result depends only on the width of the Gaussian, which must be determined numerically. Finally, to make the calculation of the ratio of determinants tractable, Ejiri expands $\log[\det M(\mu)]$ in a Taylor series in μ about $\mu = 0$. The same Taylor coefficients appear in an intermediate step in the Taylor expansion of the pressure or thermodynamic potential. Since one is expanding the action instead of the thermodynamic potential, the convergence properties are different —possibly more favorable.

Applying this method to p4fat3 staggered fermions with the Wilson gauge action, a rather coarse lattice with $N_\tau = 4$, and a rather large quark mass, Ejiri locates the critical chemical potential at $\mu/T > 2.5$, approximately. This is an interesting result, which awaits reconciliation with the questions raised by Golterman, Shamir, and Svetitsky concerning phase ambiguities of the fourth root of the staggered fermion determinant [74].

Thus we see that all of the methods, save, perhaps, the probability distribution function method, are limited to quite small chemical potentials.

5.2.7 Stochastic quantization method

All of the above lattice methods for simulating at nonzero chemical potential evaluate the Feynman path integral using Monte Carlo importance sampling, a technique that is inherently unstable when the path integrand is not positive definite. At nonzero chemical potential, the $SU(3)$ fermion determinant is complex, and the wide variety of methods outlined above deal with the complex phase with limited success. Instead of quantizing via the Feynman path integral method, Aarts and Stamatescu [100] have recently proposed using the stochastic quantization method [101]. In the early days of lattice calculations, stochastic quantization through the Langevin equation [102] was, in fact, one of the competing numerical methods for nonperturbative calculations in quantum field theory, and it met with mixed success [103].

For purposes of this review, we give just a brief sketch of stochastic quantization. For a theory with a scalar field $\phi(x)$ and action S, we generate an ensemble of fields $\phi(x,\tau)$ where τ is a fictitious Langevin time (analogous to molecular-dynamics or Markov-chain time in the standard importance sampling approach). The ensemble satisfies the stochastic equation

$$\frac{\partial \phi(x,\tau)}{\partial \tau} = -\frac{\delta S}{\delta\phi(x)} + \eta(x,\tau),$$
(59)

where $\eta(x,\tau)$ is a Gaussian random field (source), uncorrelated in x. As long as S has a well-defined minimum and we start with a solution near that minimum, without the random source the field relaxes to the classical solution where the action is stationary, *i.e.*, the variational derivative $\delta S/\delta\phi(x)$ vanishes. The random source then induces "quantum fluctuations" about the classical solution. Quantum observables are estimated in the usual way as expectation values on the equilibrium ensemble.

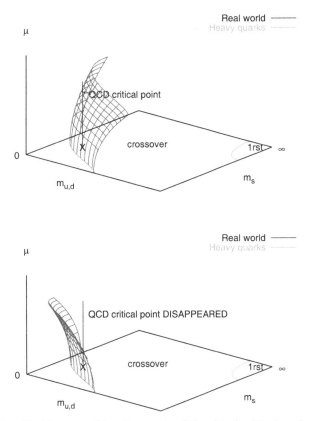

Fig. 21. Two possible alignments of the chiral critical surface at low chemical potential from [65]. Top: the scenario permitting a first-order phase transition at high densities and temperatures. Bottom: the scenario allowing only a crossover.

When the action S is complex, we get a complex solution and a complex stationary point, a region that is not reached with conventional importance sampling. The hope is that the solution is still attracted to the appropriate stationary point, *i.e.*, the Langevin method is stable. Aarts and Stamatescu have done some preliminary tests with simplified models that imitate the characteristics of QCD at nonzero chemical potential. Their results are promising [104,105].

5.3 Curvature of the critical surface

One question of considerable phenomenological importance can be addressed with simulations at small chemical potential. That is whether the $Z(2)$ critical line sketched in fig. 14 moves closer to the physical quark masses as the chemical potential is increased or it moves farther away. If it moves closer, as shown in the upper panel of fig. 21, one may expect a true phase transition in a suitably baryon-rich environment, such as may occur in a moderately low-energy heavy-ion collision. If it moves away, as shown in the lower panel, there would be no such expectation. De Forcrand and Philipsen set out to address this question using the imaginary chemical potential method. Their results at $N_\tau = 4$ suggest that the critical line moves

away [65,106–108], at least when all three quark flavors are close to having equal masses.

6 Equation of state

The equation of state gives the energy density, pressure, and/or entropy of the thermal QCD ensemble as a function of temperature at constant volume. All quantities are renormalized by subtracting their values at zero temperature. The subtraction eliminates an ultraviolet divergence, but the cancellation of this divergence makes the computation costly in the continuum limit, since one must compute the $\mathcal{O}(a^{-4})$ divergent high-temperature and zero-temperature quantities independently and subtract them to get a finite result.

There are two traditional methods for computing the equation of state and one recently introduced method.

6.1 Derivative method

The first method is based on the identity

$$\varepsilon = \frac{T^2}{V} \left. \frac{\partial \ln Z}{\partial T} \right|_V . \tag{60}$$

On the lattice the derivative with respect to temperature at fixed volume in the first identity translates to a derivative with respect to $1/(N_\tau a_t)$ at fixed a_s, where a_t is the lattice spacing in the imaginary time direction and a_s is the lattice spacing in the spatial direction. At fixed N_τ, we differentiate with respect to a_t itself.

For example, for the original Wilson plaquette gauge action of eq. (11) the explicit dependence on a_t and a_s goes as follows:

$$S_G(a_s, a_t, g^2) = 2/g^2(a_s, a_t) \left[\frac{a_s}{a_t} \sum_x P_t(x) + \frac{a_t}{a_s} \sum_x P_s(x) \right], \tag{61}$$

where we have distinguished the timelike and spacelike plaquettes

$$P_t(x) = \sum_i \operatorname{Re} \operatorname{Tr}[1 - U_{P,i,0}(x)],$$

$$P_s(x) = \sum_{i<j} \operatorname{Re} \operatorname{Tr}[1 - U_{P,i,j}(x)]. \tag{62}$$

In the gauge action above, we have indicated the dependence of the gauge coupling on the lattice constants a_s and a_t. That dependence is defined through a standard renormalization procedure for an anisotropic lattice: at a fixed ratio a_t/a_s and gauge coupling g, we compute an experimentally accessible, dimensionful quantity, such as the splitting of a quarkonium system. From the experimental value of the splitting, we can then determine the lattice constants in physical units. We repeat the procedure, varying g and a_t/a_s to get the full dependence of g on the lattice constants.

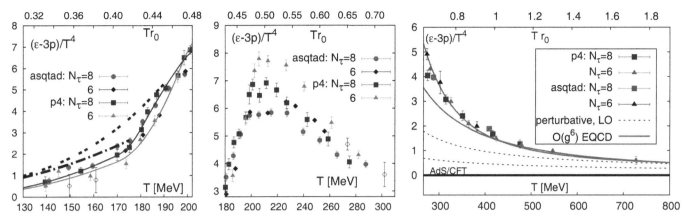

Fig. 22. Details of the dependence of the interaction measure on temperature in MeV units (bottom scale) and r_0 units (top scale) for three temperature ranges left to right: low, middle, and high, for $N_\tau = 6$ and 8 from a HotQCD study comparing p4fat3 and asqtad staggered fermion formulations [51,109]. Measurements in most cases are taken along a line of constant physics with $m_{ud} = 0.1m_s$. In the low temperature range the dashed and dash-dotted curves are predictions of a hadron resonance gas model with different high mass cutoffs. The other curves in that range are spline fits to the data. In the high-temperature range the dashed lines are the leading-order perturbative prediction for $\mu_{\overline{MS}} = 2\pi T$ and $\mu_{\overline{MS}} = \pi T$. The brown line (the line passing through the points) is a fit to leading-order perturbation theory plus a bag constant, and the magenta line (the line passing mostly below the points) is an $\mathcal{O}(g^6)$ EQCD prediction from [110]. For a brief mention of EQCD, see sect. 7.1.

So from eq. (15) with only the gauge action in this example, we have [111] (after setting $a_t = a_s = a$)

$$\varepsilon = -T \left. \frac{\partial \ln g^2}{\partial \ln a_t} \right|_{a_s} \langle S_G/V \rangle + (6/g^2)T \langle P_t - P_s \rangle. \quad (63)$$

The partial derivative of the gauge coupling with respect to a_t is called the Karsch coefficient. It is known up to 1-loop order in lattice perturbation theory, but a nonperturbative calculation described above is necessary at experimentally accessible temperatures. As we indicated above, that calculation is rather involved.

6.2 Standard integral method

A second thermodynamic identity gives the pressure as the volume derivative of the thermodynamic potential,

$$p = T \left. \frac{\partial}{\partial V} \ln Z \right|_T. \quad (64)$$

By itself, this identity leads to an expression similar to the energy density above, but in this case we need the derivative of the gauge coupling with respect to the spatial lattice spacing a_s at fixed a_t. We have the same difficulty as before in requiring a nonperturbative calculation of an unconventional quantity.

But if we combine the two identities to form the interaction measure I,

$$I = \varepsilon - 3p, \quad (65)$$

then we get a total derivative of the gauge coupling with respect to $a = a_s = a_t$ and the lattice thermodynamic identity

$$I = -\frac{T}{V}\frac{d \ln Z}{d \ln a}. \quad (66)$$

The isotropic derivative of the coupling with respect to the cutoff is just the commonly computed renormalization group beta function $\beta = dg^2/d \ln a$. For the Wilson plaquette gauge action we get

$$I = -T/V (d \ln g^2/d \ln a) \langle S_G \rangle. \quad (67)$$

So the lattice derivative is readily calculated in terms of the conventional plaquette observable and the beta function. With fermions present we require also the chiral condensate and the derivative of the quark masses with respect to the lattice spacing. These are also easily accessible in lattice calculations.

We must bear in mind that the physical quantities require subtracting the zero-temperature values, so in the end we need the difference

$$\Delta I = I(T) - I(0). \quad (68)$$

We will often drop the Δ in the following discussion and figures.

Figure 22 shows the interaction measure difference obtained in a recent $N_\tau = 8$ calculation with equal-mass up and down quarks and a strange quark. The mass of the strange quark was held fixed at approximately its physical value, and the masses of the up and down quarks were set to a fixed fraction of the strange quark mass. Thus the temperature was varied roughly along parameter space lines of constant physics, meaning light pseudoscalar mesons (at zero temperature) had approximately constant masses.

To complete the determination of the equation of state, we need the energy density and pressure separately. The pressure is easily computed in the thermodynamic limit, in which $\ln Z$ is simply proportional to the volume:

$$\ln Z = -pV/T, \quad (69)$$

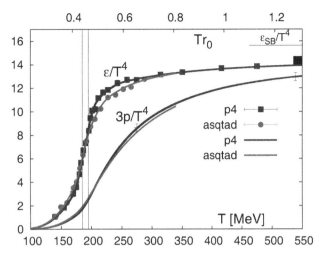

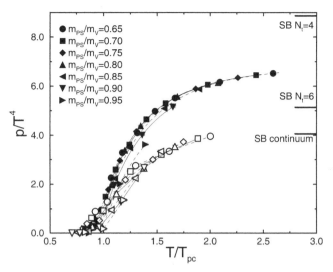

Fig. 23. Equation of state showing energy density and three times the pressure, both divided by the fourth power of the temperature *vs.* temperature for $N_\tau = 8$. Measurements are taken along a line of constant physics with $m_{ud} = 0.1m_s$. Results are from a HotQCD study comparing p4fat3 and asqtad staggered fermion formulations [51]. The blue error bars on the pressure curve indicate the size of the error. The black bar shows a systematic error from setting the lower limit of the pressure integration.

Fig. 24. The pressure as function of T/T_{pc}, with T_{pc} the pseudocritical or crossover temperature for two flavors of clover-improved Wilson fermions on $16^3 \times 4$ lattices (filled symbols) and $16^3 \times 6$ lattices (open symbols), from [112]. The simulation was done for a variety of rather heavy quark masses, indicated by the vector to pseudoscalar mass ratios m_{PS}/m_V. The lattice artifacts are larger than with improved staggered quarks, as expected from table 1.

So the expression (66) can also be written as

$$I = \frac{T}{V} \frac{\mathrm{d}(pV/T)}{\mathrm{d}\ln a}, \qquad (70)$$

or, if we fix VT^3 in the derivative, as

$$I/T^4 = \frac{\mathrm{d}(p/T^4)}{\mathrm{d}\ln T}. \qquad (71)$$

We can then use the identity (70) at fixed N_τ to integrate with respect to $\ln a$ (equivalently $\ln T$) to get the pressure:

$$p(a)a^4 - p(a_0)a_0^4 = -\int_{\ln a_0}^{\ln a} \Delta I(a')(a')^4 \, \mathrm{d}\ln a'. \qquad (72)$$

Here the lower endpoint of integration a_0 is a large lattice spacing, corresponding to a low temperature. If it is sufficiently low, we may take $p(a_0) = 0$ and the expression then yields the pressure at temperature $T = 1/(N_\tau a)$.

The integration is carried out numerically, since the integrand is determined in a series of simulations done at fixed lattice spacing. However, the spacing of the points can be set arbitrarily close as needed. The energy density is then obtained from $\varepsilon = I + 3p$ and the entropy density from $s = \varepsilon + p$.

This integral method was used to complete the construction of the equation of state with improved staggered quarks shown in fig. 23. The same method has also been used in a study with two flavors of clover improved Wilson fermions [112], as shown in fig. 24.

6.3 Temperature integral method

In the standard integral method above we fixed N_τ and integrated eq. (66) with respect to lattice spacing to get the pressure. The temperature integral method of [113] instead fixes the lattice spacing and "integrates" eq. (71) over N_τ at fixed N_s.

The advantage of working at a fixed lattice spacing (so fixed gauge coupling, quark masses, and Hamiltonian) is that the zero temperature subtraction is the same for all N_τ, and we are assured of following lines of constant physics [114]. With the standard integral method, to carry out the necessary subtraction, we need a separate zero-temperature simulation for each high-temperature point. Thus one may hope for a savings in computational effort.

The disadvantage of the temperature integral method is that the integrand is known only at the discrete temperatures $1/(N_\tau a_t)$ for integer N_τ. To decrease the sample interval at a given temperature, one must start with a smaller a_t, which increases the cost substantially. Simulating on an anisotropic lattice helps.

So far, the method has been tested on a pure Yang-Mills ensemble with the pleasing result shown in fig. 25.

6.4 Step scaling method

The standard integral method of eq. (72) has the disadvantage that it requires computing the difference between the high- and zero-temperature values of the interaction measure at each value of the gauge coupling (*i.e.*, each high-temperature point). At increasingly high temperature we get closer to the continuum limit and the matching zero-temperature calculation becomes very expensive.

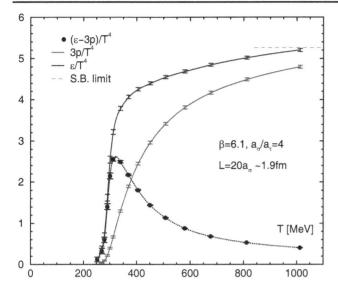

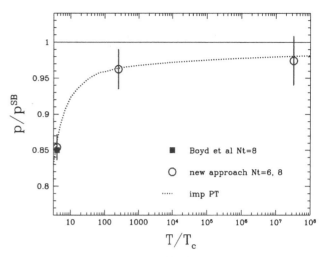

Fig. 25. Equation of state (interaction measure, energy density and pressure) for the pure Yang-Mills theory, obtained using the T integral method at fixed lattice spacing $a_\sigma = 0.097$ fm and aspect ratio $a_\sigma/a_\tau = 4$ [115,113].

Fig. 26. Circles: pressure from [116] for pure Yang-Mills theory at ultra-high temperatures compared with predictions of EQCD perturbation theory (dotted line [117,118,110]). The pressure is given in units of the Stefan-Boltzmann value and the temperature in units of the temperature at the phase transition T_c. The square is computed using the standard integral method [119].

Endrödi et al. propose a step scaling method that alleviates this problem to some degree [116]. Their idea is to compute the pressure at a given temperature as a series of differences:

$$p(T) - p(0) = [p(T) - p(T/2)] + [p(T/2) - p(T/4)] + \ldots. \tag{73}$$

The increment

$$\bar{p}(T) = p(T) - p(T/2) \tag{74}$$

must be calculated at the same cutoff a to renormalize properly the ultraviolet divergence. In practice, this means matching a calculation at a given $N_\tau = N$ with a calculation at $N_\tau = 2N$ for the same bare action parameters. (The step factor $1/2$ can be replaced by any factor less than 1.) The differences $[p(T) - p(T/2)]/T^4$ are bounded from above, so the series

$$[p(T) - p(0)]/T^4 = \bar{p}(T)/T^4 + \frac{1}{16}\bar{p}(T)/T^4|_{T/2}$$
$$+ \frac{1}{256}\bar{p}(T)/T^4|_{T/4} + \ldots \tag{75}$$

converges rapidly.

Endrödi et al. suggest two ways to calculate $\bar{p}(T)$. One uses a modified form of eq. (72):

$$p(a, N_\tau = N)a^4 - p(a, N_\tau = 2N)a^4 =$$
$$-\int_{\ln a_0}^{\ln a} [I(a', N_\tau = N) - I(a', N_\tau = 2N)](a')^4 \, d\ln a'. \tag{76}$$

Here, we have shown the N_τ-dependence explicitly. We assume that a_0 is large enough that the integration constants $p(a_0, N_\tau)$ are essentially zero.

The second method uses the identity eq. (69) to write

$$\bar{p}(T = 1/(aN)) = p(a, N_\tau = N) - p(a, N_\tau = 2N) =$$
$$[\ln Z(N_\tau = 2N) - \ln Z^2(N_\tau = N)]/(N_s^3 N). \tag{77}$$

The rhs is the difference between the partition functions on two lattices of size $N_s^3 \times 2N$ in which one lattice is intact and the other is split in half at the midpoint in imaginary time with periodic (or fermion-antiperiodic) boundary conditions applied to the two halves. To compute this difference, Endrödi et al. modify the action at the interface by introducing an interpolating parameter α such that $\alpha = 1$ corresponds to the fully split lattice and $\alpha = 0$, to the fully intact lattice. The simulation measures the derivative of $\ln Z(\alpha)$ with respect to α, which involves only fields at the interface. The increment (77) is then computed from

$$\bar{p}(T) = \frac{1}{N_s^3 N} \int_0^1 d\alpha \, \frac{d\ln Z(\alpha)}{d\alpha}. \tag{78}$$

There is still a strong cancellation involved in the integration over α, but it is a bit milder than the cancellation in the standard integral method. With their method they are able to reach such high temperatures that contact with perturbation theory is certainly expected, as shown in fig. 26. For a lower temperature comparison of the $O(g^6)$ EQCD prediction of Laine et al. [110] with the interaction measure computed using standard methods, see fig. 22. For a brief mention of EQCD, see sect. 7.1.

6.5 Equation of state at nonzero densities

Heavy-ion collisions involve interacting hadronic matter at relatively low baryon densities and high temperatures. At the other extreme, high baryon densities and low temperatures may occur in the cores of dense stars. In both cases we would like to know the equation of state. For the low-density environment of heavy-ion collisions the Taylor series method is effective for lattice simulations. Unfortunately, thus far we have no reliable lattice method to simulate the conditions of dense stars.

Consider the $(2+1)$-flavor case of equal nonzero up- and down-quark chemical potentials $\mu_u = \mu_d = \mu_{ud}$ and a nonzero strange chemical potential μ_s. The pressure can be expanded as follows:

$$\frac{p}{T^4} = \sum_{n,m=0}^{\infty} c_{nm}(T) \left(\frac{\mu_{ud}}{T}\right)^n \left(\frac{\mu_s}{T}\right)^m, \quad (79)$$

The coefficients c_{nm} are evaluated at zero chemical potential

$$c_{nm}(T) = \frac{1}{n!}\frac{1}{m!}\frac{1}{T^3V}\frac{\partial^{n+m}\ln Z}{\partial(\mu_{ud}/T)^n\partial(\mu_s/T)^m}\bigg|_{\mu_{ud,s}=0}. \quad (80)$$

CP symmetry requires that the coefficients vanish for odd $n+m$ at zero chemical potential.

For increasing n and m the coefficients c_{nm} are increasingly complicated combinations of traces of the inverse of the lattice Dirac matrix. For a simple example, the lowest-order mixed coefficient is

$$c_{11} = \left\langle \text{Tr}\left(M_{ud}^{-1}\frac{\partial M_{ud}}{\partial(\mu_{ud}/T)}\right) \text{Tr}\left(M_s^{-1}\frac{\partial M_s}{\partial(\mu_s/T)}\right)\right\rangle. \quad (81)$$

Such observables are technically difficult to compute because the trace is over all lattice sites as well as over colors. Usually such traces are evaluated by stochastic sampling methods. As the order n and m increase, not only are the traces more complicated, the required number of stochastic samples grows rapidly. In effect, the computational effort grows factorially in the expansion order.

The quark number densities $\langle n_{ud}\rangle$ and $\langle n_s\rangle$ can be found from first derivatives in the same expansion. For $\langle n_{ud}\rangle$ it is

$$\langle n_{ud}\rangle = \frac{1}{V}\frac{\partial \ln Z}{\partial(\mu_{ud}/T)} =$$

$$T^3 \sum_{n=1,m=0}^{\infty} n c_{nm}(T) \left(\frac{\mu_{ud}}{T}\right)^{n-1}\left(\frac{\mu_s}{T}\right)^m, \quad (82)$$

and for $\langle n_s\rangle$,

$$\langle n_s\rangle = \frac{1}{V}\frac{\partial \ln Z}{\partial(\mu_s/T)} =$$

$$T^3 \sum_{n=0,m=1}^{\infty} m c_{nm}(T) \left(\frac{\mu_{ud}}{T}\right)^n \left(\frac{\mu_s}{T}\right)^{m-1}. \quad (83)$$

The leading terms in the expansion are

$$\frac{\langle n_s\rangle}{T^3} \approx c_{11}(T)\left(\frac{\mu_{ud}}{T}\right) + c_{02}(T)\left(\frac{\mu_s}{T}\right). \quad (84)$$

The mixed coefficient $c_{11}(T)$ is nonzero (and negative) at low temperatures, because when we add a strange quark

to the ensemble, it is screened by a light antiquark. This tendency persists at temperatures close to, but above the crossover. So for $\mu_{ud} \neq 0$, the strange quark number density is nonzero for $\mu_s = 0$. In heavy-ion collisions the mean strange quark number density is zero, so we need to "tune" the strange quark chemical potential to obtain the experimental conditions.

The quark number susceptibility matrix χ_{ab} for $a,b \in u,d,s$ is likewise found from second derivatives. For example, for the diagonal elements and the equivalent mixed light off-diagonal elements $\chi_{uu} = \chi_{dd} = \chi_{ud} = \chi_{du}$, we have

$$\chi_{uu} = \frac{\partial\langle n_{ud}/T\rangle}{\partial(\mu_{ud}/T)} =$$

$$T^2 \sum_{n=2,m=0}^{\infty} n(n-1)c_{nm}(T)\left(\frac{\mu_{ud}}{T}\right)^{n-2}\left(\frac{\mu_s}{T}\right)^m. (85)$$

The (diagonal) strange quark number susceptibility $\chi_s = \chi_{ss}$ is similarly obtained. The heavy-light mixed quark number susceptibility $\chi_{us} = \chi_{su}$ is

$$\chi_{us} = \frac{\partial\langle n_{ud}/T\rangle}{\partial(\mu_s/T)} =$$

$$T^2 \sum_{n=1,m=1}^{\infty} nm c_{nm}(T)\left(\frac{\mu_{ud}}{T}\right)^{n-1}\left(\frac{\mu_s}{T}\right)^{m-1}. (86)$$

The interaction measure can also be expanded in this way [120]. Once we have both pressure and interaction measure, we can determine the energy density and entropy density for any small chemical potential. As an example, we show the equation of state at constant entropy density per baryon number in fig. 27. This is the equation of state appropriate to an adiabatic expansion or compression of hadronic matter, conditions that may obtain in a heavy-ion collision.

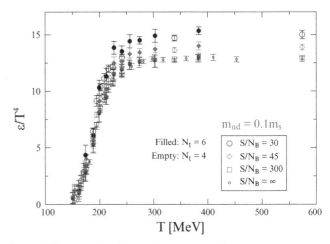

Fig. 27. Energy density *vs.* temperature for constant entropy per baryon number, from [121].

7 In-medium properties of hadrons

7.1 Spatial string tension

Despite its popular characterization as deconfined, high-temperature hadronic matter retains vestiges of confinement. Spacelike Wilson loops still exhibit the area law behavior associated with confinement. This is readily seen by considering dimensional reduction, in which for $T \gg T_c$, the short Euclidean time dimension (of extent $1/T$) is collapsed, leaving three spatial dimensions [122,123]. Since all dimensions are Euclidean, any one of them can be interpreted as Euclidean "time." We do a 90° rotation to turn one of the original spatial coordinates into the Euclidean time coordinate of a $(2 + 1)$-dimensional field theory.

The reduction of 4d QCD to what is sometimes called "EQCD" [117] has these characteristics:

- Quarks acquire a large 3d mass $\sqrt{(\pi T)^2 + m_q^2}$. This happens because the antiperiodic boundary condition in the small dimension requires a minimum momentum component πT for that coordinate, which then contributes to the energy-momentum relation as an additional effective mass.
- The original fourth component of the color vector potential A_0 is reinterpreted as a scalar Higgs-like field. The other three vector potential components become the usual vector potential of the $(2 + 1)$-dimensional theory. We get a confining gauge-Higgs theory.
- The 3d and 4d gauge couplings are related through $g_3 = g_4\sqrt{T}$.
- The spatial Wilson loop of the original 4d theory is now interpreted as the standard space-time–oriented Wilson loop of the 3d theory. Because the theory is still confining in 3d, we get a linearly rising potential with a string tension.

In a recent calculation Cheng et al. compared the behavior of the spatial string tension of the full 4d theory with predictions based on a perturbative connection between the four- and three-dimensional coupling and the numerically measured proportionality between string tension and coupling in three-dimensional $SU(3)$ Yang-Mills theory [124]. The comparison is shown in fig. 28. The good agreement at temperatures as low as $1.5T_c$ is unexpected.

7.2 Screening masses

The Yukawa potential can be thought of as a measure of the spatial correlation of a pion source and sink (the sources and sinks being static nucleons). The important insight here is that the screening mass m_π is the mass of a propagating particle. In the high-temperature plasma we can consider similar correlations between interpolating operators of any type. These spatial correlators are controlled by confined states, as we indicated in sect. 7.1. Because we no longer have Lorentz invariance, the spatial screening masses are not expected to be equal to frequencies of real-time plasma excitations, but one can speculate

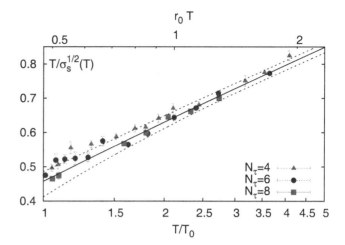

Fig. 28. Temperature divided by the square root of the spatial string tension σ_s vs. temperature in units of the crossover temperature T_0 (lower scale) and in r_0 units (upper scale) for $2+1$ flavors of p4fat3 quarks on lattices with $N_\tau = 4$, 6 and 8. The solid curve (with uncertainties indicated by the dashed lines) is the prediction of the dimensionally reduced theory [124].

that there may be a connection [125]. In any case, they provide information about the structure of the plasma, they control the behavior of a variety of susceptibilities, and their degeneracy patterns provide information about the temperature dependence of symmetries.

Euclidean thermal hadron propagators (correlators) are defined in the same way as they are at zero temperature:

$$C_{AB}(x) = \langle O_A(x) O_B(0)\rangle, \qquad (87)$$

where $O_A(x)$ and $O_B(x)$ are interpolating operators for the desired hadronic state.

At zero temperature it is typical to project the correlator to zero spatial momentum, resulting in a time-slice correlator

$$C_{AB}(t) = \int \mathrm{d}^3\mathbf{x}\, C_{AB}(t,\mathbf{x}). \qquad (88)$$

At large Euclidean time such a correlator has the asymptotic behavior

$$C_{AB}(t) \sim Z_A Z_B \exp(-Mt), \qquad (89)$$

where M is the mass of the hadron and Z_A and Z_B are overlap constants.

At nonzero temperatures one cannot explore the asymptotic limit because of the bound on Euclidean time $0 \le t \le 1/T$, but one can define a spatial correlator by fixing one of the spatial coordinates and integrating over the other three, as in

$$C_{AB}(z) = \int \mathrm{d}t\, \mathrm{d}x\, \mathrm{d}y\, C_{AB}(t,x,y,z). \qquad (90)$$

(For fermions, it is necessary to include a Matsubara phase factor $\exp[i\pi Tt]$.) For large z the asymptotic behavior is

$$C_{AB}(z) \sim Z_A(T) Z_B(T) \exp[-\mu(T)z], \qquad (91)$$

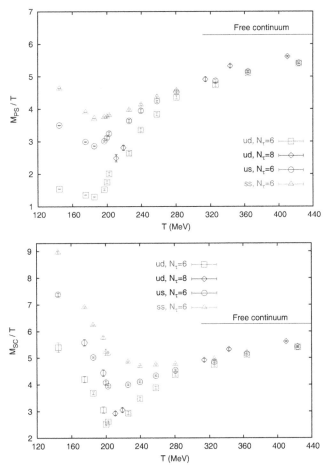

Fig. 29. Screening masses for the pseudoscalar channel (upper panel) and scalar channel (lower panel) *vs.* temperature in a dynamical $(2+1)$-flavor simulation with p4fat3 staggered fermions [126]. Measurements were taken along lines of constant physics with $m_\pi \sim 220\,\mathrm{MeV}$, $m_K = 500\,\mathrm{MeV}$ and $N_\tau = 6$ and 8 [127].

where $\mu(T)$ is the hadronic screening mass. At zero temperature $\mu(T=0) = M$.

Even though the high-temperature plasma exhibits deconfining characteristics in its real time behavior, the spatial correlations remain confined, so the spectrum of spatial meson and baryon screening masses retains a gap characteristic of confinement even in the high-temperature plasma. However, since the screening mass for quarks approach πT at high temperatures, the valence-quark-antiquark meson screening masses approach $2\pi T$ and the valence-three-quark baryon screening masses approach $3\pi T$. Furthermore, as chiral symmetry is approximately restored at high temperatures, they must exhibit the approximate degeneracies required by the chiral multiplets.

Armed with this background let us consider the temperature behavior of the screening mass $\mu_\pi(T)$ of the pion. At low temperatures the pion is a Goldstone boson, so the screening mass is small. Above the transition chiral symmetry is restored. So the screening mass rises above the transition temperature, approaching $2\pi T$. The transition temperature is marked by the change of slope. Figure 29 illustrates this behavior.

The isosinglet scalar f_0 (σ) meson can be generated using the isosinglet chiral condensate $\bar\psi\psi_{\mathrm{sing}} = \bar\psi\psi_u + \bar\psi\psi_d$ as the interpolating operator. It has a sizable mass at low temperature, but it joins the pion chiral multiplet at the transition temperature when the pion screening mass is quite small. Thus its mass must dip at the transition temperature and rise again, approaching $2\pi T$. Thus a dip in μ_{f_0} also marks the transition temperature.

The chiral susceptibilities are related to hadron propagators in Euclidean space-time. For example, the isosinglet chiral susceptibility is

$$\chi_{\mathrm{sing}}(T) = \int \mathrm{d}^4 x \, \langle \bar\psi\psi_{\mathrm{sing}}(x)\bar\psi\psi_{\mathrm{sing}}(0)\rangle =$$
$$\int \mathrm{d}z \, C_{\mathrm{sing}}(z,T), \qquad (92)$$

where $C_{\mathrm{sing}}(z)$ is the scalar-isosinglet screening correlator generated by the isosinglet chiral condensate. In addition to the f_0, this correlator also contains a two-pion continuum contribution. So its asymptotic behavior has terms in $\exp(-\mu_{f_0}z)$ as well as $\exp(-2E_\pi z)$ for $E_\pi \geq \mu_\pi$. Integration over z of these asymptotic terms yields contributions to the susceptibility that go as the inverse of the screening masses. At low temperatures the two-pion threshold is below the f_0, so the two-pion continuum dominates the susceptibility. In the chiral limit this contribution is responsible for the $1/\sqrt{m}$ singularity in the susceptibility. At high temperatures the pion screening mass rises, and the f_0 screening mass is approximately degenerate with it. Thus the two-pion continuum is expected to have a higher screening mass than the f_0, and the susceptibility is finite in the chiral limit. Thus this susceptibility should be large at low temperatures and fall abruptly at the transition temperature.

7.3 Charmonium

To the extent the transition to a quark-gluon plasma is a crossover and not a genuine phase transition, one should not expect low temperature properties to change abruptly at the crossover temperature. Confined hadronic states may persist as plasma excitations at least for temperatures close to, but above the crossover temperature. One of the most studied examples is the J/ψ, since it is readily observed experimentally, and, because of their large mass, charmed quarks are a good theoretical probe. Numerical simulation suggests that the J/ψ persists to temperatures as high as $1.5T_c$ [128,129]. (See sect. 7.3.3 below.) As the temperature increases beyond T_c, it is thought that screening of the heavy-quark potential eventually prevents the formation of a bound state and J/ψ production is suppressed [130,129].

7.3.1 Static quark/antiquark free energy

There are two lattice methods for studying thermal effects in quarkonium. The first, more model-dependent method, is based on a Born-Oppenheimer approximation [130]. One measures the free energy of a static quark-antiquark pair as a function of separation r. The result is introduced into the Schrödinger equation as a temperature-dependent potential $V(r, T)$ for a given heavy quark mass. As the temperature increases, screening effects weaken the potential, and eventually it does not support a bound state for quarks of the given mass. This approximation should be good, provided the Born-Oppenheimer adiabatic approximation is good, i.e., as long as the plasma is able to relax to its equilibrium state on the time scale of the orbital motion of the quarks.

Gauge invariance presents a subtlety in fashionable methods for extracting the free energy to be used as a Born-Oppenheimer potential. It is popular to distinguish between color-singlet and color-octet states of the static quark and antiquark. Since those states are supposed to be defined in terms of the colors of only the spatially separated quarks themselves, the separation is gauge dependent and probably not phenomenologically significant [131].

The potential method can be tested entirely in the context of a lattice calculation. One starts from the lattice static potential, derives the spectral function for the thermal quarkonium propagator (see the next subsection), and compares the result with a direct determination of the lattice spectral function. If the static approximation is correct, the results should agree. Recent attempts to follow this approach for $T_c < T < 1.5T_c$ fail to reproduce any charmonium states in the spectral function nor any but the $1S$ state of bottomonium [132]. So is the determination of the lattice spectral function unreliable, or is the static approximation unreliable for charmonium, or are both unreliable?

Related attempts have been made to derive a heavy-quark potential suitable for use in the Schrödinger equation in real time (as opposed to lattice imaginary time), but so far the methodology is developed only in perturbation theory [133–135].

7.3.2 Spectral density

The second method is model independent, but more difficult. One measures the spectral function of a thermal Green's function for the J/ψ [136]. The correlator is defined for some suitable local interpolating operator $\mathcal{O}(x_0, \mathbf{x})$ as

$$C(x_0, \mathbf{x}, T) = \langle \mathcal{O}(x_0, \mathbf{x}) \mathcal{O}(0, 0) \rangle. \qquad (93)$$

The spectral density $\rho(\omega, \mathbf{q}, T)$ is then obtained by inverting the Kubo formula for the partial Fourier transform $C(x_0, \mathbf{q}, T)$ of the correlator:

$$C(x_0, \mathbf{q}, T) = \frac{1}{2\pi} \int_0^\infty d\omega\, \rho(\omega, \mathbf{q}, T) K(\omega, x_0, T), \qquad (94)$$

where

$$K(\omega, x_0, T) = \frac{\cosh \omega(x_0 - 1/2T)}{\sinh(\omega/2T)}. \qquad (95)$$

Going from the Euclidean correlator $C(x_0, \mathbf{q}, T)$ to the spectral density $\rho(\omega, \mathbf{q}, T)$ is a very difficult inverse problem. One would like to extract detailed information about the spectral density from quite limited information. Because of time-reflection symmetry, a simulation at $N_\tau = 8$ has only five, typically noisy, independent values.

Possible remedies include 1) assuming a functional form for ρ and fitting its parameters (e.g., a delta function for the J/ψ or a Breit-Wigner shape), 2) decreasing the time interval a_t, allowing a larger N_τ, and 3) adding further constraints on ρ, as in the maximum entropy method. We outline the last remedy in the next subsection.

7.3.3 Maximum entropy method

The maximum entropy method has been used to determine spectral functions in condensed matter physics for some time [137]. It was first applied to lattice QCD by Asakawa, Hatsuda, and Nakamura [138,139]. It is essentially a Bayesian method with a prior inspired by Occam's razor. One begins by defining an unremarkable default prior spectral density $\rho_0(\omega, T)$. A typical choice would be the spectral density of a noninteracting quark-antiquark pair, or at least the density expected at asymptotically high frequency. One then requires that the spectral density ρ, inferred from the correlator data, should deviate only as much from ρ_0 as the data seems to require.

The method is applied in the context of a maximum likelihood fit to the correlator data. We give a simplified description of the method. Starting from a parameterization of the spectral density $\rho(\omega, T)$, one predicts the correlator data and computes the usual chisquare $\chi^2[\rho]$ difference between prediction and data. One introduces a Shannon-Jaynes entropy for this ρ as follows:

$$S[\rho] = \int_0^\infty d\omega\, [\rho(\omega) - \rho_0(\omega) - \rho(\omega) \ln[\rho(\omega)/\rho_0(\omega)]]. \qquad (96)$$

The "entropy" vanishes when $\rho = \rho_0$ and for small deviations from ρ_0, it is

$$S[\rho] \approx -\frac{1}{2} \int_0^\infty d\omega\, [\rho(\omega) - \rho_0(\omega)]^2/\rho_0(\omega). \qquad (97)$$

So the default prior maximizes the entropy. One then maximizes the likelihood $\exp(Q[\alpha, \rho])$ or, equivalently, $Q[\alpha, \rho]$ itself:

$$Q[\alpha, \rho] = \alpha S[\rho] - \chi^2[\rho]/2. \qquad (98)$$

The positive weight α controls the balance between maximum entropy and minimum chisquare. In the "state-of-the-art" method, the mean of the best fits $\bar{\rho}$ is then obtained from the average:

$$\bar{\rho} = \int d[\rho] d\alpha\, \exp(Q[\alpha, \rho]). \qquad (99)$$

This is our answer for the spectral density.

This method was used by Asakawa and Hatsuda to study the fate of charmonium in the high-temperature

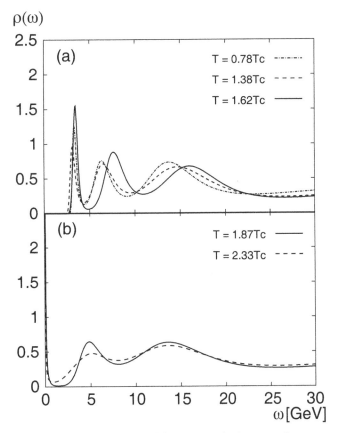

Fig. 30. Spectral density $\rho(\omega)$ for the J/ψ for several temperatures shown in units of the crossover temperature T_c [128]. The ground-state peak is visible up to $1.62T_c$. These results are obtained in a quenched simulation.

medium [128]. See also [129] and, more recently, [140]. Their results for the J/ψ spectral density are shown in fig. 30 and provided some of the first evidence that the J/ψ exists as a discernible plasma resonance for temperatures at least as high as $1.62T_c$ before it "melts."

When data are inadequate, results of the MEM method can be quite sensitive to the choice of the default model. For example, one may obtain artifact excited-state peaks. For some examples, see [141].

8 Transport coefficients

8.1 Shear and bulk viscosities

Among the transport coefficients, the shear and bulk viscosities are essential to the hydrodynamical modeling of the expansion and cooling of the quark-gluon plasma in the aftermath of a heavy-ion collision. They are obtained from correlators of the energy momentum tensor at temperature T

$$C_{\mu\nu,\rho\sigma}(x_0, \mathbf{x}, T) = \langle T_{\mu\nu}(x_0, \mathbf{x}) T_{\rho\sigma}(0) \rangle . \quad (100)$$

We need its spectral function ρ, which we obtain from its partial Fourier transform $C_{\mu\nu,\rho\sigma}(x_0, \mathbf{q}, T)$ and the Kubo

formula

$$C_{\mu\nu,\rho\sigma}(x_0, \mathbf{q}, T) = \int_0^\infty d\omega \, \rho_{\mu\nu,\rho\sigma}(\omega, \mathbf{q}, T) K(\omega, x_0, T), \quad (101)$$

where $K(\omega, x_0, T)$ is given by eq. (95). The shear (η) and bulk (ζ) viscosities are obtained from the low-frequency behavior of the spectral function $\rho(\omega, \mathbf{q}, T)$:

$$\eta(T) = \pi \lim_{\omega \to 0} \frac{\rho_{12,12}(\omega, 0, T)}{\omega} ,$$

$$\zeta(T) = \frac{\pi}{9} \lim_{\omega \to 0} \frac{\rho_{ii,jj}(\omega, 0, T)}{\omega} . \quad (102)$$

Computing the viscosity has been a well-known challenging problem since it was first attempted by Karsch and Wyld [142]. The correlator is noisy, requiring high statistics. As with the J/ψ correlator, this is a difficult inverse problem. A further complication is that the spectral function has a nasty T-independent, large ω, ultraviolet behavior $\rho \sim \omega^4$, which tends to overwhelm the low-frequency contribution to $C(x, \tau)$ for low x_0.

Possible remedies include 1) assuming a functional form for ρ and fitting its parameters [142], 2) decreasing the time interval a_t, allowing a larger N_τ [143], and 3) adding further constraints on ρ, such as maximum entropy [144], and working at small nonzero momentum [145].

Meyer [146–148] has done a new high-statistics calculation in pure Yang-Mills theory and uses a parameterization of the spectral function in terms of an optimized basis set that folds in appropriate perturbative behavior at large ω and then emphasizes deviations from this behavior. For the ratio of shear viscosity to entropy density, he finds $\eta/s = 0.134(33)$ at $1.65T_c$ where perturbation theory gives 0.8, and for the ratio of bulk viscosity to entropy density, $\zeta/s < 0.15$ at $1.65T_c$ and $\zeta/s < 0.015$ at $3.2T_c$. These results support the notion that the plasma is a nearly perfect fluid.

8.2 Dilepton emission and related quantities

The dilepton emission rate, the soft photon emissivity, and the electrical conductivity of the plasma are other important transport properties. They are obtained from the thermal correlator of the electric current

$$G_{EM}(x_0, \mathbf{x}, T) = \langle J_\mu(x_0, \mathbf{x}) J_\mu(0) \rangle , \quad (103)$$

$$G_{EM}(x_0, \mathbf{q}, T) = \int_0^\infty \frac{d\omega}{2\pi} K(\omega, x_0, T) \rho_{EM}(\omega, \mathbf{q}, T). \quad (104)$$

Again, this is a difficult inverse problem. The ultraviolet divergence is milder here than with the spectral function of the stress energy tensor. In this case $\rho \sim \omega^2$. Otherwise, the same methods have been applied.

The spectral density $\rho_{EM}(\omega, 0, T)$ determines the differential dilepton pair production rate [149]:

$$\left. \frac{dW^4}{d\omega d^3 p} \right|_{\mathbf{p}=0} = \frac{5\alpha_{rmem}^2}{27\pi^2} \frac{1}{\omega^2(e^{\omega/T} - 1)} \rho_{EM}(\omega, 0, T). \quad (105)$$

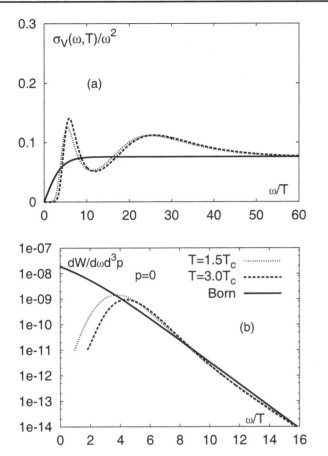

Fig. 31. Relationship between (a) the vector meson spectral density $\rho_{EM}(\omega, 0, T)$ (shown here as $\sigma_V(\omega, T)$) and (b) the dilepton differential production rate $dW/d\omega d^3p$ at zero three-momentum, plotted as a function of energy ω in units of temperature for two temperatures above the crossover temperature T_c [150]. The solid lines represent a free quark-antiquark pair. The dashed and dotted lines are lattice MEM results that show a peak corresponding to a vector meson resonance. Results are obtained in the quenched approximation.

An example of the relationship between the MEM determination of the spectral function and the resulting dilepton rate is given by Karsch *et al.* [150] in fig. 31. These results show a strong enhancement over the free quark-antiquark pair contribution, at least up to three times T_c resulting from a vector meson resonance. The hard dilepton rate is obtained from the spectral function for $\omega/T \gg 1$, and there is rough agreement between perturbation theory and lattice simulation.

As with the shear and bulk viscosity, the challenge is getting to low frequency to obtain the soft photon emissivity, and at zero frequency, the electrical conductivity:

$$\sigma(T) = \frac{1}{16} \frac{\partial}{\partial \omega} \rho_{EM}(\omega, \mathbf{0}, T)\bigg|_{\omega=0}, \qquad (106)$$

Extracting the spectral function itself is challenging enough. Extracting its derivative compounds the difficulty. Gupta *et al.* [151] tried different Bayesian priors to constrain the spectral function.

9 Outlook

Numerical simulations have taught us much about the properties of high-temperature strongly interacting matter. Here are highlights discussed in this article:

- We have a fair understanding of the QCD phase diagram at nonzero temperature and zero or small baryon densities and nearly physical quark masses.
- We have a phenomenologically useful determination of the equation of state.
- We have a good understanding of the behavior of the quark number susceptibility.
- We are beginning to understand the small mass limit of the chiral condensate and its related susceptibilities.
- We know the plasma has persistent confining properties that are observable in screening masses and the spatial string tension.
- We have some indications of the persistence of hadronic states as resonances in the plasma phase at temperatures close to and above T_c.
- We are starting to determine plasma transport coefficients.
- We are starting to make contact with perturbation theory at high temperatures.

There are many outstanding questions. Here are particularly pressing ones:

- We need a more robust determination of transport coefficients.
- We do not have a good way to simulate at moderately large or higher nonzero baryon number densities.
- We do not know, yet, whether the critical point in the $(\mu/T, T)$-plane is experimentally accessible.
- We do not know whether the tricritical point in the $(m_s, m_{u,d})$-plane lies above or below the physical m_s.
- We would like to understand better the behavior of the equation of state in the region where it overlaps with hadron resonance gas models.
- It would be good to develop more confidence in our understanding of the continuum limit of phenomenologically important quantities.
- It would be good to have high precision results from fermion formulations other than staggered for purposes of corroboration.
- We would like to develop more confidence in our contact with perturbation theory at high temperatures.

Work currently underway will help resolve some of these issues. At zero or small baryon number densities we expect progress with Wilson quark formulations, including clover-improved and twisted-mass. Simulations with domain-wall quarks will help test conclusions about chiral properties. Forthcoming simulations with highly improved staggered quarks (HISQ) will help reduce some of the lattice artifacts of the staggered fermion formulation, especially at temperatures leading up to T_c, where we suspect they are important.

For simulations at nonzero baryon number densities we really need some new ideas. Perhaps stochastic quantization will help. For transport coefficients and the small-quark-mass region of the phase diagram, we may expect progress simply by applying more computing power.

Lattice QCD thermodynamics is a very active field. We expect continued strong progress in the years to come.

We thank Ludmila Levkova for a careful reading of the manuscript. This work is supported by the National Science Foundation under grants PHY04-56691 and PHY07-57333.

References

1. A. Bazavov et al., Full nonperturbative QCD simulations with 2+1 flavors of improved staggered quarks, arXiv: 0903.3598.

2. F. Wilczek, QCD in extreme conditions, arXiv: hep-ph/0003183.

3. R.P. Feynman, A.R. Hibbs, Quantum Mechanics and Path Integrals (McGraw-Hill, New York, 1965).

4. I. Montvay, G. Münster, Quantum Fields on the Computer (Cambridge University Press, Cambridge, 1997).

5. M. Creutz, Quantum Fields on the Computer (World Scientific, Singapore, 1992).

6. T. DeGrand, C. DeTar, Lattice Methods for Quantum Chromodynamics (World Scientific, Singapore, 2006).

7. K.G. Wilson, Confinement of quarks, Phys. Rev. D 10, 2445 (1974).

8. A.M. Wegner, Duality in generalized Ising models and phase transitions without local order parameters, J. Math. Phys. 12, 2259 (1971).

9. A.M. Polyakov, Compact gauge fields and the infrared catastrophe, Phys. Lett. B 59, 82 (1975).

10. M. Lüscher, P. Weisz, On-Shell Improved Lattice Gauge Theories, Commun. Math. Phys. 97, 59 (1985).

11. M. Lüscher, P. Weisz, Computation of the Action for On-Shell Improved Lattice Gauge Theories at Weak Coupling, Phys. Lett. B 158, 250 (1985).

12. K. Symanzik, Continuum Limit and Improved Action in Lattice Theories. 2. O(N) Nonlinear Sigma Model in Perturbation Theory, Nucl. Phys. B 226, 205 (1983).

13. B. Sheikholeslami, R. Wohlert, Improved Continuum Limit Lattice Action for QCD with Wilson Fermions, Nucl. Phys. B 259, 572 (1985).

14. Alpha Collaboration (R. Frezzotti, P.A. Grassi, S. Sint, P. Weisz), Lattice QCD with a chirally twisted mass term, JHEP 08, 058 (2001) [arXiv: hep-lat/0101001].

15. R. Frezzotti, G.C. Rossi, Chirally improving Wilson fermions. I: O(a) improvement, JHEP 08, 007 (2004) [arXiv: hep-lat/0306014].

16. CP-PACS Collaboration (A. Ali Khan et al.), Phase structure and critical temperature of two flavor QCD with renormalization group improved gauge action and clover improved Wilson quark action, Phys. Rev. D 63, 034502 (2001) [arXiv: hep-lat/0008011].

17. S. Ejiri, Lattice QCD thermodynamics with Wilson quarks, Prog. Theor. Phys. Suppl. 168, 245 (2007) [arXiv: 0704.3747].

18. HPQCD Collaboration (E. Follana, A. Hart, C.T.H. Davies, Q. Mason), The low-lying Dirac spectrum of staggered quarks, Phys. Rev. D 72, 054501 (2005) [arXiv: hep-lat/0507011].

19. S.R. Sharpe, Rooted staggered fermions: Good, bad or ugly?, PoS LAT2006, 022 (2006) [arXiv: hep-lat/0610094].

20. P. Lepage, Perturbative improvement for lattice QCD: An update, Nucl. Phys. B Proc. Suppl. 60, 267 (1998) [arXiv: hep-lat/9707026].

21. MILC Collaboration (C.W. Bernard et al.), Quenched hadron spectroscopy with improved staggered quark action, Phys. Rev. D 58, 014503 (1998) [arXiv: hep-lat/9712010].

22. J.F. Lagae, D.K. Sinclair, Improving the staggered quark action to reduce flavour symmetry violations, Nucl. Phys. Proc. Suppl. 63, 892 (1998) [arXiv: hep-lat/9709035].

23. J.F. Lagae, D.K. Sinclair, Improved staggered quark actions with reduced flavour symmetry violations for lattice QCD, Phys. Rev. D 59, 014511 (1999) [arXiv: hep-lat/9806014].

24. MILC Collaboration (K. Orginos, D. Toussaint), Testing improved actions for dynamical Kogut-Susskind quarks, Phys. Rev. D 59, 014501 (1999) [arXiv: hep-lat/9805009].

25. MILC Collaboration (D. Toussaint, K. Orginos), Tests of improved Kogut-Susskind fermion actions, Nucl. Phys. Proc. Suppl. 73, 909 (1999) [arXiv: hep-lat/9809148].

26. G.P. Lepage, Flavor-symmetry restoration and Symanzik improvement for staggered quarks, Phys. Rev. D 59, 074502 (1999) [arXiv: hep-lat/9809157].

27. MILC Collaboration (K. Orginos, D. Toussaint, R.L. Sugar), Variants of fattening and flavor symmetry restoration, Phys. Rev. D 60, 054503 (1999) [arXiv: hep-lat/9903032].

28. MILC Collaboration (C.W. Bernard et al.), Scaling tests of the improved Kogut-Susskind quark action, Phys. Rev. D 61, 111502 (2000) [arXiv: hep-lat/9912018].

29. U.M. Heller, F. Karsch, B. Sturm, Improved staggered fermion actions for QCD thermodynamics, Phys. Rev. D 60, 114502 (1999) [arXiv: hep-lat/9901010].

30. F. Karsch, E. Laermann, A. Peikert, The pressure in 2, 2+1 and 3 flavour QCD, Phys. Lett. B 478, 447 (2000) [arXiv: hep-lat/0002003].

31. HPQCD Collaboration (E. Follana et al.), Highly improved staggered quarks on the lattice, with applications to charm physics, Phys. Rev. D 75, 054502 (2007) [arXiv: hep-lat/0610092].

32. Y. Aoki, Z. Fodor, S.D. Katz, K.K. Szabó, The QCD transition temperature: Results with physical masses in the continuum limit, Phys. Lett. B 643, 46 (2006) [arXiv: hep-lat/0609068].

33. N. Ishizuka, M. Fukugita, H. Mino, M. Okawa, A. Ukawa, Operator dependence of hadron masses for Kogut-Susskind quarks on the lattice, Nucl. Phys. B 411, 875 (1994).

34. D.B. Kaplan, A Method for simulating chiral fermions on the lattice, Phys. Lett. B 288, 342 (1992) [arXiv: hep-lat/9206013].

35. V. Furman, Y. Shamir, Axial symmetries in lattice QCD with Kaplan fermions, Nucl. Phys. B 439, 54 (1995) [arXiv: hep-lat/9405004].

36. R.G. Edwards, U.M. Heller, R. Narayanan, Spectral flow, chiral condensate and topology in lattice QCD, Nucl. Phys. B 535, 403 (1998) [arXiv: hep-lat/9802016].

37. M. Golterman, Y. Shamir, B. Svetitsky, Localization properties of lattice fermions with plaquette and improved gauge actions, Phys. Rev. D 72, 034501 (2005) [arXiv: hep-lat/0503037].

38. RBC Collaboration (D.J. Antonio *et al.*), *Localization and chiral symmetry in 3 flavor domain wall QCD*, Phys. Rev. D **77**, 014509 (2008) [arXiv: 0705.2340].

39. P. Chen *et al.*, *The finite temperature QCD phase transition with domain wall fermions*, Phys. Rev. D **64**, 014503 (2001) [arXiv: hep-lat/0006010].

40. RBC-HotQCD Collaboration (M. Cheng), *QCD Thermodynamics with Domain Wall Fermions*, PoS LAT2008, 180 (2008) [arXiv: 0810.1311].

41. R. Narayanan, H. Neuberger, *A Construction of lattice chiral gauge theories*, Nucl. Phys. B **443**, 305 (1995) [arXiv: hep-th/9411108].

42. H. Neuberger, *Exactly massless quarks on the lattice*, Phys. Lett. B **417**, 141 (1998) [arXiv: hep-lat/9707022].

43. P.H. Ginsparg, K.G. Wilson, *A Remnant of Chiral Symmetry on the Lattice*, Phys. Rev. D **25**, 2649 (1982).

44. M. Lüscher, *Exact chiral symmetry on the lattice and the Ginsparg-Wilson relation*, Phys. Lett. B **428**, 342 (1998) [arXiv: hep-lat/9802011].

45. Z. Fodor, S.D. Katz, K.K. Szabó, *Dynamical overlap fermions, results with hybrid Monte-Carlo algorithm*, JHEP **08**, 003 (2004) [arXiv: hep-lat/0311010].

46. P. Hegde, F. Karsch, E. Laermann, S. Shcheredin, *Lattice cut-off effects and their reduction in studies of QCD thermodynamics at non-zero temperature and chemical potential*, Eur. Phys. J. C **55**, 423 (2008) [arXiv: 0801.4883].

47. W. Bietenholz, R. Brower, S. Chandrasekharan, U.J. Wiese, *Progress on perfect lattice actions for QCD*, Nucl. Phys. Proc. Suppl. **53**, 921 (1997) [arXiv: hep-lat/9608068].

48. R. Sommer, *A New way to set the energy scale in lattice gauge theories and its applications to the static force and alpha-s in SU(2) Yang-Mills theory*, Nucl. Phys. B **411**, 839 (1994) [arXiv: hep-lat/9310022].

49. R. Gupta *et al.*, *The EOS from simulations on BlueGene L Supercomputer at LLNL and NYBlue*, PoS LAT2008, 170 (2008).

50. HotQCD Collaboration (C. DeTar, R. Gupta), *Toward a precise determination of T_c with 2+1 flavors of quarks*, PoS LAT2007, 179 (2007) [arXiv: 0710.1655].

51. HotQCD Collaboration (A. Bazavov *et al.*), *Equation of state and QCD transition at finite temperature*, arXiv: 0903.4379.

52. MILC Collaboration (C. Bernard *et al.*), *QCD thermodynamics with three flavors of improved staggered quarks*, Phys. Rev. D **71**, 034504 (2005) [arXiv: hep-lat/0405029].

53. M. Peskin, D.V. Schroeder, *An Introduction to Quantum Field Theory* (Westview Press, New York, 1995).

54. RBC-Bielefeld Collaboration, F. Karsch, *Fluctuations of Goldstone modes and the chiral transition in QCD*, Nucl. Phys. A **820**, 99C (2008) [arXiv: 0810.3078].

55. Y. Maezawa *et al.*, *Thermodynamics and heavy-quark free energies at finite temperature and density with two flavors of improved Wilson quarks*, PoS LAT2007, 207 (2007) [arXiv: 0710.0945].

56. W. Söldner, *Quark Mass Dependence of the QCD Equation of State on $N_\tau = 8$ Lattices*, PoS LAT2008, 173 (2008).

57. Y. Aoki *et al.*, *The QCD transition temperature: results with physical masses in the continuum limit II*, arXiv: 0903.4155.

58. MILC Collaboration (C. Aubin *et al.*), *Light hadrons with improved staggered quarks: Approaching the continuum limit*, Phys. Rev. D **70**, 094505 (2004) [arXiv: hep-lat/0402030].

59. G. Boyd *et al.*, *Equation of state for the SU(3) gauge theory*, Phys. Rev. Lett. **75**, 4169 (1995) [arXiv: hep-lat/9506025].

60. R.D. Pisarski, F. Wilczek, *Remarks on the Chiral Phase Transition in Chromodynamics*, Phys. Rev. D **29**, 338 (1984).

61. F. Karsch, E. Laermann, C. Schmidt, *The chiral critical point in 3-flavor QCD*, Phys. Lett. B **520**, 41 (2001) [arXiv: hep-lat/0107020].

62. F. Karsch *et al.*, *Where is the chiral critical point in 3-flavor QCD?*, Nucl. Phys. Proc. Suppl. **129**, 614 (2004) [arXiv: hep-lat/0309116].

63. Y. Aoki, G. Endrödi, Z. Fodor, S.D. Katz, K.K. Szabó, *The order of the quantum chromodynamics transition predicted by the standard model of particle physics*, Nature **443**, 675 (2006) [arXiv: hep-lat/0611014].

64. E. Laermann, O. Philipsen, *Status of lattice QCD at finite temperature*, Annu. Rev. Nucl. Part. Sci. **53**, 163 (2003) [arXiv: hep-ph/0303042].

65. P. de Forcrand, O. Philipsen, *The chiral critical line of $N_f = 2+1$ QCD at zero and non-zero baryon density*, JHEP **01**, 077 (2007) [arXiv: hep-lat/0607017].

66. T. Mendes, *Universality and Scaling at the chiral transition in two-flavor QCD at finite temperature*, PoS LAT2007, 208 (2007) [arXiv: 0710.0746].

67. J.B. Kogut, D.K. Sinclair, *Evidence for O(2) universality at the finite temperature transition for lattice QCD with 2 flavours of massless staggered quarks*, Phys. Rev. D **73**, 074512 (2006) [arXiv: hep-lat/0603021].

68. M. D'Elia, A. Di Giacomo, C. Pica, *Two flavor QCD and confinement*, Phys. Rev. D **72**, 114510 (2005) [arXiv: hep-lat/0503030].

69. A.M. Halasz, A.D. Jackson, R.E. Shrock, M.A. Stephanov, J.J.M. Verbaarschot, *On the phase diagram of QCD*, Phys. Rev. D **58**, 096007 (1998) [arXiv: hep-ph/9804290].

70. F. Karsch, *Lattice simulations of the thermodynamics of strongly interacting elementary particles and the exploration of new phases of matter in relativistic heavy ion collisions*, J. Phys. Conf. Ser. **46**, 122 (2006) [arXiv: hep-lat/0608003].

71. J.B. Kogut *et al.*, *Chiral Symmetry Restoration in Baryon Rich Environments*, Nucl. Phys. B **225**, 93 (1983).

72. M.G. Alford, K. Rajagopal, F. Wilczek, *Color-flavor locking and chiral symmetry breaking in high density QCD*, Nucl. Phys. B **537**, 443 (1999) [arXiv: hep-ph/9804403].

73. R. Rapp, T. Schafer, E.V. Shuryak, M. Velkovsky, *Diquark Bose condensates in high density matter and instantons*, Phys. Rev. Lett. **81**, 53 (1998) [arXiv: hep-ph/9711396].

74. M. Golterman, Y. Shamir, B. Svetitsky, *Breakdown of staggered fermions at nonzero chemical potential*, Phys. Rev. D **74**, 071501 (2006) [arXiv: hep-lat/0602026].

75. O. Philipsen, *The QCD phase diagram at zero and small baryon density*, PoS LAT2005, 016 (2006) [arXiv: hep-lat/0510077].

76. S. Ejiri, *Recent progress in lattice QCD at finite density*, PoS LAT2008, 002 (2008).

77. A.M. Ferrenberg, R.H. Swendsen, *New Monte Carlo Technique for Studying Phase Transitions*, Phys. Rev. Lett. **61**, 2635 (1988).

78. A.M. Ferrenberg, R.H. Swendsen, *Optimized Monte Carlo analysis*, Phys. Rev. Lett. **63**, 1195 (1989).

79. D. Toussaint, *Simulating QCD at finite density*, Nucl. Phys. Proc. Suppl. **17**, 248 (1990).

80. Z. Fodor, S.D. Katz, *A new method to study lattice QCD at finite temperature and chemical potential*, Phys. Lett. B **534**, 87 (2002) [arXiv: hep-lat/0104001].

81. Z. Fodor, S.D. Katz, *Critical point of QCD at finite T and μ, lattice results for physical quark masses*, JHEP **04**, 050 (2004) [arXiv: hep-lat/0402006].

82. Z. Fodor, S.D. Katz, *Lattice determination of the critical point of QCD at finite T and μ*, JHEP **03**, 014 (2002) [arXiv: hep-lat/0106002].

83. J.B. Kogut, D.K. Sinclair, *Lattice QCD at finite isospin density at zero and finite temperature*, Phys. Rev. D **66**, 034505 (2002) [arXiv: hep-lat/0202028].

84. J.B. Kogut, D.K. Sinclair, *Lattice QCD at finite temperature and density in the phase-quenched approximation*, Phys. Rev. D **77**, 114503 (2008) [arXiv: 0712.2625].

85. P. de Forcrand, M.A. Stephanov, U. Wenger, *On the phase diagram of QCD at finite isospin density*, PoS LAT2007, 237 (2007) [arXiv: hep-lat/0711.0023].

86. I.M. Barbour, S.E. Morrison, E.G. Klepfish, J.B. Kogut, M.-P. Lombardo, *Results on finite density QCD*, Nucl. Phys. B Proc. Suppl. **60**, 220 (1998) [arXiv: hep-lat/9705042].

87. J. Engels, O. Kaczmarek, F. Karsch, E. Laermann, *The quenched limit of lattice QCD at non-zero baryon number*, Nucl. Phys. B **558**, 307 (1999) [arXiv: hep-lat/9903030].

88. P. de Forcrand, S. Kratochvila, *Finite density QCD with a canonical approach*, Nucl. Phys. Proc. Suppl. **153**, 62 (2006) [arXiv: hep-lat/0602024].

89. A. Alexandru, M. Faber, I. Horvath, K.-F. Liu, *Lattice QCD at finite density via a new canonical approach*, Phys. Rev. D **72**, 114513 (2005) [arXiv: hep-lat/0507020].

90. X.-f. Meng, A. Li, A. Alexandru, K.-F. Liu, *Winding number expansion for the canonical approach to finite density simulations*, PoS LATTICE2008, 032 (2008) [arXiv: 0811.2112].

91. A. Li, X. Meng, A. Alexandru, K.-F. Liu, *Finite Density Simulations with Canonical Ensemble*, PoS LATTICE2008, 178 (2008) [arXiv: 0810.2349].

92. P. de Forcrand, O. Philipsen, *The QCD phase diagram for small densities from imaginary chemical potential*, Nucl. Phys. B **642**, 290 (2002) [arXiv: hep-lat/0205016].

93. M. D'Elia, M.-P. Lombardo, *Finite density QCD via imaginary chemical potential*, Phys. Rev. D **67**, 014505 (2003) [arXiv: hep-lat/0209146].

94. C.R. Allton et al., *The QCD thermal phase transition in the presence of a small chemical potential*, Phys. Rev. D **66**, 074507 (2002) [arXiv: hep-lat/0204010].

95. R.V. Gavai, S. Gupta, *Pressure and non-linear susceptibilities in QCD at finite chemical potentials*, Phys. Rev. D **68**, 034506 (2003) [arXiv: hep-lat/0303013].

96. R.V. Gavai, S. Gupta, *The critical end point of QCD*, Phys. Rev. D **71**, 114014 (2005) [arXiv: hep-lat/0412035].

97. R.V. Gavai, S. Gupta, *QCD at finite chemical potential with six time slices*, Phys. Rev. D **78**, 114503 (2008) [arXiv: 0806.2233].

98. Z. Fodor, S.D. Katz, C. Schmidt, *The density of states method at non-zero chemical potential*, JHEP **03**, 121 (2007) [arXiv: hep-lat/0701022].

99. S. Ejiri, *On the existence of the critical point in finite density lattice QCD*, Phys. Rev. D **77**, 014508 (2008) [arXiv: 0706.3549].

100. G. Aarts, I.-O. Stamatescu, *Stochastic quantization at finite chemical potential*, JHEP **09**, 018 (2008) [arXiv: 0807.1597].

101. G. Parisi, Y.-s. Wu, *Perturbation Theory Without Gauge Fixing*, Sci. Sin. **24**, 483 (1981).

102. G. Parisi, *On complex probabilities*, Phys. Lett. B **131**, 393 (1983).

103. P.H. Damgaard, H. Hüffel, *Stochastic Quantization*, Phys. Rep. **152**, 227 (1987).

104. G. Aarts, *Can stochastic quantization evade the sign problem? – the relativistic Bose gas at finite chemical potential*, Phys. Rev. Lett. **102**, 131601 (2009) [arXiv: 0810.2089].

105. G. Aarts, *Complex Langevin dynamics at finite chemical potential: mean field analysis in the relativistic Bose gas*, JHEP **05**, 052 (2009) [arXiv: 0902.4686].

106. P. de Forcrand, S. Kim, O. Philipsen, *A QCD chiral critical point at small chemical potential: is it there or not?*, PoS LAT2007, 178 (2007) [arXiv: 0711.0262].

107. P. de Forcrand, O. Philipsen, *The curvature of the critical surface $(m_{ud}, m_s)^{crit}(\mu)$: a progress report*, PoS LATTICE2008, 208 (2008) [arXiv: 0811.3858].

108. P. de Forcrand, O. Philipsen, *The chiral critical point of $N_f = 3$ QCD at finite density to the order $(\mu/T)^4$*, JHEP **11**, 012 (2008) [arXiv: 0808.1096].

109. P. Petreczky, *Quark Matter 2009: Finite temperature lattice QCD: Present status*.

110. M. Laine, Y. Schröder, *Quark mass thresholds in QCD thermodynamics*, Phys. Rev. D **73**, 085009 (2006) [arXiv: hep-ph/0603048].

111. J. Engels, J. Fingberg, F. Karsch, D. Miller, M. Weber, *Nonperturbative thermodynamics of SU(N) gauge theories*, Phys. Lett. B **252**, 625 (1990).

112. CP-PACS Collaboration (A. Ali Khan et al.), *Equation of state in finite-temperature QCD with two flavors of improved Wilson quarks*, Phys. Rev. D **64**, 074510 (2001) [arXiv: hep-lat/0103028].

113. T. Umeda et al., *Fixed Scale Approach to Equation of State in Lattice QCD*, Phys. Rev. D **79**, 051501 (2009) [arXiv: 0809.2842].

114. L. Levkova, T. Manke, R. Mawhinney, *Two-flavor QCD thermodynamics using anisotropic lattices*, Phys. Rev. D **73**, 074504 (2006) [arXiv: hep-lat/0603031].

115. T. Umeda et al., *Thermodynamics of SU(3) gauge theory at fixed lattice spacing*, PoS LAT2008, 174 (2008).

116. G. Endrödi, Z. Fodor, S.D. Katz, K.K. Szabó, *The equation of state at high temperatures from lattice QCD*, PoS LAT2007, 228 (2007) [arXiv: 0710.4197].

117. E. Braaten, A. Nieto, *Free Energy of QCD at High Temperature*, Phys. Rev. D **53**, 3421 (1996) [arXiv: hep-ph/9510408].

118. K. Kajantie, M. Laine, K. Rummukainen, Y. Schröder, *The pressure of hot QCD up to $g^6 \ln(1/g)$*, Phys. Rev. D **67**, 105008 (2003) [arXiv: hep-ph/0211321].

119. G. Boyd et al., *Thermodynamics of SU(3) Lattice Gauge Theory*, Nucl. Phys. B **469**, 419 (1996) [arXiv: hep-lat/9602007].

120. MILC Collaboration (C. Bernard *et al.*), *QCD thermo-dynamics with 2+1 flavors at nonzero chemical potential*, Phys. Rev. D **77**, 014503 (2008) [arXiv: 0710.1330].

121. MILC Collaboration (S. Basak *et al.*), *QCD equation of state at non-zero chemical potential*, PoS LAT2008, 171 (2008).

122. P.H. Ginsparg, *First Order and Second Order Phase Transitions in Gauge Theories at Finite Temperature*, Nucl. Phys. B **170**, 388 (1980).

123. T. Appelquist, R.D. Pisarski, *High-Temperature Yang-Mills Theories and Three-Dimensional Quantum Chromodynamics*, Phys. Rev. D **23**, 2305 (1981).

124. M. Cheng *et al.*, *The Spatial String Tension and Dimensional Reduction in QCD*, Phys. Rev. D **78**, 034506 (2008) [arXiv: 0806.3264].

125. C.E. DeTar, *A Conjecture Concerning the Modes of Excitation of the Quark-Gluon Plasma*, Phys. Rev. D **32**, 276 (1985).

126. M. Cheng *et al.*, *The QCD Equation of State with almost Physical Quark Masses*, Phys. Rev. D **77**, 014511 (2008) [arXiv: 0710.0354].

127. E. Laermann *et al.*, *Recent results on screening masses*, PoS LAT2008, 193 (2008).

128. M. Asakawa, T. Hatsuda, *J/ψ and η_c in the deconfined plasma from lattice QCD*, Phys. Rev. Lett. **92**, 012001 (2004) [arXiv: hep-lat/0308034].

129. S. Datta, F. Karsch, P. Petreczky, I. Wetzorke, *Behavior of charmonium systems after deconfinement*, Phys. Rev. D **69**, 094507 (2004) [arXiv: hep-lat/0312037].

130. T. Matsui, H. Satz, *J/ψ Suppression by Quark-Gluon Plasma Formation*, Phys. Lett. B **178**, 416 (1986).

131. O. Jahn, O. Philipsen, *The Polyakov loop and its relation to static quark potentials and free energies*, Phys. Rev. D **70**, 074504 (2004) [arXiv: hep-lat/0407042].

132. A. Mocsy, P. Petreczky, *Can quarkonia survive deconfinement?*, Phys. Rev. D **77**, 014501 (2008) [arXiv: 0705.2559].

133. M. Laine, O. Philipsen, P. Romatschke, M. Tassler, *Real-time static potential in hot QCD*, JHEP **03**, 054 (2007) [arXiv: hep-ph/0611300].

134. M.A. Escobedo, J. Soto, *Non-relativistic bound states at finite temperature (I): the hydrogen atom*, Phys. Rev. A **78**, 032520 (2009) [arXiv: 0804.0691].

135. N. Brambilla, J. Ghiglieri, A. Vairo, P. Petreczky, *Static quark-antiquark pairs at finite temperature*, Phys. Rev. D **78**, 014017 (2008) [arXiv: 0804.0993].

136. T. Umeda, K. Nomura, H. Matsufuru, *Charmonium at finite temperature in quenched lattice QCD*, Eur. Phys. J. C **39**, s1, 9 (2005) [arXiv: hep-lat/0211003].

137. M. Jarrell, J. Gubernatis, *Bayesian inference and the analytic continuation of imaginary-time quantum Monte Carlo data*, Phys. Rep. **269**, 133 (1996).

138. Y. Nakahara, M. Asakawa, T. Hatsuda, *Hadronic spectral functions in lattice QCD*, Phys. Rev. D **60**, 091503 (1999) [arXiv: hep-lat/9905034].

139. M. Asakawa, T. Hatsuda, Y. Nakahara, *Maximum entropy analysis of the spectral functions in lattice QCD*, Prog. Part. Nucl. Phys. **46**, 459 (2001) [arXiv: hep-lat/0011040].

140. G. Aarts, C. Allton, M.B. Oktay, M. Peardon, J.-I. Skullerud, *Charmonium at high temperature in two-flavor QCD*, Phys. Rev. D **76**, 094513 (2007) [arXiv: 0705.2198].

141. H.T. Ding, O. Kaczmarek, F. Karsch, H. Satz, *Charmonium correlators and spectral functions at finite temperature*, PoS (Confinement8), 108 (2008) [arXiv: 0901.3023].

142. F. Karsch, H.W. Wyld, *Thermal Green's functions and transport coefficients on the lattice*, Phys. Rev. D **35**, 2518 (1987).

143. G. Burgers, F. Karsch, A. Nakamura, I.O. Stamatescu, *QCD on anisotropic lattices*, Nucl. Phys. B **304**, 587 (1988).

144. G. Aarts, C. Allton, J. Foley, S. Hands, S. Kim, *Spectral functions at small energies and the electrical conductivity in hot, quenched lattice QCD*, Phys. Rev. Lett. **99**, 022002 (2007) [arXiv: hep-lat/0703008].

145. H.B. Meyer, *Energy-momentum tensor correlators and spectral functions*, JHEP **08**, 031 (2008) [arXiv: 0806.3914].

146. H. Meyer, *Energy-momentum tensor correlators and viscosity*, PoS LAT2008, 017 (2008).

147. H.B. Meyer, *A calculation of the bulk viscosity in SU(3) gluodynamics*, Phys. Rev. Lett. **100**, 162001 (2008) [arXiv: 0710.3717].

148. H.B. Meyer, *A calculation of the shear viscosity in SU(3) gluodynamics*, Phys. Rev. D **76**, 101701 (2007) [arXiv: 0704.1801].

149. E. Braaten, R.D. Pisarski, T.-C. Yuan, *Production of soft dileptons in the quark - gluon plasma*, Phys. Rev. Lett. **64**, 2242 (1990).

150. F. Karsch, E. Laermann, P. Petreczky, S. Stickan, I. Wetzorke, *A lattice calculation of thermal dilepton rates*, Phys. Lett. B **530**, 147 (2002) [arXiv: hep-lat/0110208].

151. S. Gupta, *The electrical conductivity and soft photon emissivity of the QCD plasma*, Phys. Lett. B **597**, 57 (2004) [arXiv: hep-lat/0301006].

Hadronic parity violation and effective field theory

B.R. Holstein[a]

Department of Physics-LGRT, University of Massachusetts, Amherst, MA 01003, USA

Received: 23 March 2009 / Revised: 14 May 2009
Published online: 23 June 2009 – © Società Italiana di Fisica / Springer-Verlag 2009
Communicated by U.-G. Meißner

Abstract. The history and phenomenology of hadronic parity violation is reviewed and a new model-independent approach based on effective field theory is developed. Possible future developments are discussed.

PACS. 11.30.Er Charge conjugation, parity, time reversal, and other discrete symmetries – 21.30.Fe Forces in hadronic systems and effective interactions – 25.20.-x Photonuclear reactions – 25.40.Lw Radiative capture

Contents

1 Introduction

The strong parity-conserving nucleon-nucleon interaction has, of course, been well studied since the beginning of quantum mechanics. Indeed, the cornerstone of traditional nuclear physics is the study of the nuclear force and, over the years, phenomenological forms of the nuclear potential have become increasingly sophisticated. In the nucleon-nucleon (NN) system, where data abound, the present state of the art is indicated, for example, by phenomenological potentials such as AV18 that are able to fit phase shifts in the energy region from threshold to 350 MeV in terms of ~ 40 parameters. Great progress has also been made in the description of few-nucleon systems [1].

At the same time, in recent years a new technique —effective field theory (EFT)— has been used in order to attack this problem using the symmetries of QCD [2].

In this approach the nuclear interaction is separated into long- and short-distance components. In its original formulation [3], designed for processes with typical momenta comparable to the pion mass —$Q \sim m_\pi$— the long-distance component is described fully quantum mechanically in terms of pion exchange, while the short-distance piece is described in terms of a small number of phenomenologically determined contact couplings. The resulting potential [4,5] is approaching [6,7] the degree of accuracy of purely phenomenological potentials. Even higher precision can be achieved at lower momenta —$Q \ll m_\pi$— where all interactions can be taken as short-ranged, as has been demonstrated not only in the NN system [8,9], but also in the three-nucleon system [10,11]. Precise —$\sim 1\%$— values have been generated for low-energy, astrophysically important cross-sections for reactions such as $n + p \to d + \gamma$ [12] and $p + p \to d + e^+ + \nu_e$ [13]. However, besides providing reliable values for such quantities, the use of EFT techniques allows for a realistic estimation of the size of possible corrections.

Because of the presence of the weak interactions, there exists, of course, in addition to the parity-conserving strong force, a parity-*violating NN* interaction, the study of which began in 1957 with an experiment by Tanner seeking (but not finding) parity violation in the $^{19}\mathrm{F}(p,\alpha)^{16}\mathrm{O}$ reaction [14]. Since that time there have been numerous additional experiments involving both nucleons and nuclei as well as considerable theoretical work. However, despite more than a half century of effort, there remain considerable problems in understanding this weak hadronic PV interaction. The first systematic theoretical basis for understanding this interaction was a pion exchange plus local interaction picture posited by Blin-Stoyle in 1960 [15]. The local interaction piece was developed into a vector-meson-exchange term by Michel in

---

[a] e-mail: `holstein@physics.umass.edu`

1964 [16]. Then in 1980 this approach was developed into a comprehensive theoretical framework by Desplanques, Donoghue, and Holstein (DDH) [17]. The latter is the basis of the analysis of nearly all experimental work which has been done during the past quarter century.

As will be discussed in more detail below, the goal of this work has been to measure the phenomenological weak parity-violating meson-nucleon coupling constants defined by DDH. However, in spite of a great deal of effort on this problem there is still no way to describe all the experimental results in terms of the DDH picture. Of particular interest is the size of the weak πNN coupling constant, where there is disagreement as to whether it is of the same general size or is considerably smaller than the value (gu)estimated by DDH. In order to sort out whether the problems in analyzing such experiments are due to the model-dependent meson exchange picture used by DDH or on account of some deeper issue, in recent years Zhu *et al.* have developed a systematic effective field theory approach to study of the PV NN interaction, and that is the subject which is outlined below [18].

Since EFT methods are somewhat unfamiliar to some physicists, in the next section we contrast the conventional and EFT approach to study of the familiar low-energy parity-conserving NN interactions, with and without Coulomb effects, and demonstrate via either approach that near-threshold observables are expressible in terms of just *two* phenomenological parameters —the singlet and triplet S-wave scattering lengths. Then, we show how such EFT methods can be extended to the PV NN interaction at the cost of introducing *five* new parameters —the Danilov coefficients [19]. In sect. 3 we indicate how these parameters are related to the underlying effective PV NN Lagrangian and in sect. 4 we describe how they can be extracted from experiments on light nuclei. (We emphasize the use of *light* nuclei in order to ameliorate nuclear-physics uncertainties.) We conclude our paper with a look into the future, when such a program is reality, and suggest new directions for work at that time. We close with a brief recapitulation.

2 Parity-conserving NN scattering

We begin our discussion with a brief look at conventional scattering theory [20], where in the usual partial-wave expansion, we can write the scattering amplitude as

$$f(\theta) = \sum_\ell (2\ell + 1) a_\ell(k) P_\ell(\cos \theta). \tag{1}$$

Here the partial-wave amplitude $a_\ell(k)$ has the form

$$a_\ell(k) = \frac{1}{k} e^{i\delta(k)} \sin \delta(k) = \frac{1}{k \cot \delta(k) - ik}. \tag{2}$$

Below we contrast the conventional and EFT approaches to the analysis of this scattering process. We begin with the conventional potential model technique.

2.1 Conventional analysis

Working in the potential model picture, one specifies a potential $V(r)$ describing the interaction of two particles, taken for simplicity here to be spinless, yielding a general expression for the scattering phase shift $\delta_\ell(k)$

$$\sin \delta_\ell(k) = -k \int_0^\infty dr' r' j_\ell(kr') 2m_r V(r') u_{\ell,k}(r'), \tag{3}$$

where m_r is the reduced mass and

$$\begin{aligned} u_{\ell,k}(r) = {} & r \cos \delta_\ell(k) j_\ell(kr) \\ & + kr \int_0^r dr' r' j_\ell(kr') n_\ell(kr) u_{\ell,k}(r') 2m_r V(r') \\ & + kr \int_r^\infty dr' r' j_\ell(kr) n_\ell(kr') u_{\ell,k}(r') 2m_r V(r') \end{aligned} \tag{4}$$

is the scattering wave function [20]. At very low energies one can characterize the analytic function $k^{2\ell+1} \cot \delta(k)$ via the effective range expansion [21]

$$k^{2\ell+1} \cot \delta_\ell(k) = -\frac{1}{a_\ell} + \frac{1}{2} r_\ell^e k^2 + \dots . \tag{5}$$

Then, from eq. (3) we can calculate *all* of the effective range parameters —e.g., in the case of a weak potential the scattering length a_ℓ is given by

$$a_\ell = \frac{1}{[(2\ell+1)!!]^2} \int_0^\infty dr'(r')^{2\ell+2} 2m_r V(r') + \mathcal{O}(V^2). \tag{6}$$

As a specific example of the use of potential methods, suppose we utilize a simple square well potential to describe the interaction

$$V(r) = \begin{cases} -V_0, & r \le R, \\ 0, & r > R. \end{cases} \tag{7}$$

For the S-wave scattering the wave function in the interior and exterior regions can then be written as

$$\psi^{(+)}(r) = \begin{cases} N j_0(Kr), & r \le R, \\ N' e^{i\delta_0}(j_0(kr) \cos \delta_0 - n_0(kr) \sin \delta_0), & r > R, \end{cases} \tag{8}$$

where $j_0(kr), n_0(kr)$ are spherical Bessel functions and the interior, exterior wave numbers are given by $k = \sqrt{2m_r E}$, $K = \sqrt{2m_r(E + V_0)}$, respectively. The connection between the two forms can be made by matching logarithmic derivatives at the boundary, yielding

$$k \cot \delta \simeq -\frac{1}{R}\left[1 + \frac{1}{KRF(KR)}\right] \quad \text{with} \quad F(x) = \cot x - \frac{1}{x}. \tag{9}$$

Making the effective range expansion —eq. (5)— we can find expressions for the scattering length, effective range, and higher moments. Thus, defining $K_0 = \sqrt{2m_r V_0}$

$$a_0 = R\left[1 - \frac{\tan K_0 R}{K_0 R}\right],$$

$$r_0^e = -\frac{1}{K_0}\left[\frac{K_0 R \sec^2 K_0 R - \tan K_0 R}{(K_0 R - \tan K_0 R)^2}\right], \quad \text{etc.} \tag{10}$$

Note that for weak potentials —$K_0 R \ll 1$— this expression for the scattering length agrees with the general result, eq. (6)

$$a_0 = \int_0^\infty \mathrm{d}r' r'^2 2m_r V(r') = -\frac{2m_r}{3} R^3 V_0 + \mathcal{O}(V_0^2). \quad (11)$$

The important feature here is that because we have chosen a specific form of the potential, *all* terms in the effective range expansion are predicted. Of course, the forms given above in the case of weak potentials are modified in the general case, but the entire system of parameters is determined to all orders and can be determined numerically.

Our application of this formalism will be to the two-nucleon system, so that we must also introduce spin degrees of freedom. We note then that at very low energies, where only the scattering length is relevant, we can write the S-wave scattering matrix in the phenomenological form [22]

$$\mathcal{M}_{PC}(\boldsymbol{k}', \boldsymbol{k}) = m_t(k) P_1 + m_s(k) P_0, \quad (12)$$

where

$$P_1 = \frac{1}{4}(3 + \boldsymbol{\sigma}_1 \cdot \boldsymbol{\sigma}_2), \qquad P_0 = \frac{1}{4}(1 - \boldsymbol{\sigma}_1 \cdot \boldsymbol{\sigma}_2)$$

are spin-triplet, -singlet spin projection operators and

$$m_t(k) \simeq \frac{-a_t}{1 + ika_t}, \qquad m_s(k) \simeq \frac{-a_s}{1 + ika_s} \quad (13)$$

are the S-wave partial-wave amplitudes in the lowest-order effective range approximation, keeping only the scattering lengths a_t, a_s. Here the scattering cross-section is found via

$$\frac{\mathrm{d}\sigma}{\mathrm{d}\Omega} = \mathrm{Tr}\,\mathcal{M}^\dagger \mathcal{M}, \quad (14)$$

so that we reproduce the familiar result

$$\frac{\mathrm{d}\sigma_{s,t}}{\mathrm{d}\Omega} = \frac{|a_{s,t}|^2}{1 + k^2 a_{s,t}^2}. \quad (15)$$

The corresponding scattering wave functions are then given by

$$\psi_{\boldsymbol{k}}^{(+)}(\boldsymbol{r}) = \left[e^{i\boldsymbol{k}\cdot\boldsymbol{r}} - \frac{M}{4\pi} \int \mathrm{d}^3 r' \frac{e^{ik|\boldsymbol{r}-\boldsymbol{r}'|}}{|\boldsymbol{r}-\boldsymbol{r}'|} U(\boldsymbol{r}') \psi_{\boldsymbol{k}}^{(+)}(\boldsymbol{r}) \right] \chi$$

$$\xrightarrow{r\to\infty} \left[e^{i\boldsymbol{k}\cdot\boldsymbol{r}} + \mathcal{M}(-i\boldsymbol{\nabla}, \boldsymbol{k}) \frac{e^{ikr}}{r} \right] \chi, \quad (16)$$

where χ is the spin function. In Born approximation then we can write the scattering wave function in terms of an effective delta function potential

$$U_{t,s}(\boldsymbol{r}) = \frac{4\pi}{M}(a_t P_1 + a_s P_0)\delta^3(\boldsymbol{r}) \quad (17)$$

as can be confirmed by substitution into eq. (16). (Strictly speaking, the use of the Born approximation is not legitimate for the case of singular potentials such as we use,

and must be properly defined as in [23].) Of course, before applying this result we need to enforce the stricture of unitarity, which requires that

$$2\,\mathrm{Im}\,T = T^\dagger T. \quad (18)$$

In the case of the S-wave partial-wave amplitude $m_t(k)$ this condition reads

$$\mathrm{Im}\,m_t(k) = k|m_t(k)|^2 \quad (19)$$

and requires the form

$$m_t(k) = \frac{1}{k} e^{i\delta_t(k)} \sin \delta_t(k). \quad (20)$$

Since at zero energy we have

$$\lim_{k\to 0} m_t(k) = -a_t, \quad (21)$$

it is clear that unitarity can be enforced by modifying this lowest-order result via

$$m_t(k) = \frac{-a_t}{1 + ika_t}, \quad (22)$$

which is simply the lowest-order effective range result, so everything seems to hang together.

For the case of nucleon-nucleon interactions, one finds[1]

$$a_0^s = -23.715 \pm 0.015\,\mathrm{fm}, \qquad r_0^s = 2.73 \pm 0.03\,\mathrm{fm},$$
$$a_0^t = 5.423 \pm 0.005\,\mathrm{fm}, \qquad r_0^t = 1.73 \pm 0.02\,\mathrm{fm}, \quad (23)$$

for scattering in the spin-singlet ($S = 0$) and spin-triplet ($S = 1$) channels, respectively. The existence of a bound state with energy $E = -\gamma^2/2m_r$ is indicated by the presence of a pole along the positive-imaginary k-axis —i.e. $\gamma > 0$ under the analytic continuation $k \to i\gamma$—

$$\frac{1}{a_0} + \frac{1}{2} r_0 \gamma^2 - \gamma = 0. \quad (24)$$

We see from eq. (23) that there exists no bound state in the np spin-singlet channel, but in the spin-triplet system there exists a solution

$$\kappa = \frac{1 - \sqrt{1 - \frac{2r_0^t}{a_0^t}}}{r_0^t} = 45.7\,\mathrm{MeV}, \quad i.e. \quad E_B = -2.23\,\mathrm{MeV} \quad (25)$$

corresponding to the deuteron.

2.2 Coulomb effects

When Coulomb interactions are included, the analysis becomes more challenging but remains straightforward. Suppose first that only same charge (*e.g.*, proton-proton) scattering is considered and that we describe the interaction in terms of a potential of the form

$$V(r) = \begin{cases} U(r), & r < R, \\ \frac{\alpha}{r}, & r > R, \end{cases} \quad (26)$$

---

[1] Note that the large scattering lengths found here show that this is certainly *not* a weak potential situation.

i.e. an attractive component —$U(r)$— at short distances, in order to mimic the strong interaction, and the repulsive Coulomb potential —α/r— at large distance, where $\alpha \simeq 1/137$ is the fine-structure constant. The analysis of scattering then proceeds as above but with the replacement of the exterior spherical Bessel functions j_0, n_0 by the corresponding Coulomb wave functions F_0^+, G_0^+

$$j_0(kr) \rightarrow F_0^+(r), \qquad n_0(kr) \rightarrow G_0^+(r), \qquad (27)$$

whose explicit form can be found in ref. [24]. For our purposes we require only the form of these functions in the limit $kr \ll 1$:

$$F_0^+(r) \xrightarrow{kr\ll 1} C(\eta_+(k)) \left(1 + \frac{r}{2a_B} + \ldots \right)$$

$$G_0^+(r) \xrightarrow{kr\ll 1} -\frac{1}{C(\eta_+(k))} \left\{ \frac{1}{kr} \right.$$
$$\left. + 2\eta_+(k) \left[h(\eta_+(k)) + 2\gamma_E - 1 + \ln \frac{r}{a_B} \right] + \ldots \right\}.$$
$$(28)$$

Here $\gamma_E = 0.577215\ldots$ is Euler's constant,

$$C^2(\eta_+(k)) = \frac{2\pi\eta_+(k)}{\exp(2\pi\eta_+(k)) - 1} \equiv K_s \qquad (29)$$

is the usual Coulombic enhancement factor, $a_B = 1/m_r\alpha$ is the Bohr radius, $\eta_+(k) = 1/ka_B$, and

$$h(\eta_+(k)) = \operatorname{Re} H(i\eta_+(k))$$
$$= \eta_+^2(k) \sum_{n=1}^{\infty} \frac{1}{n(n^2 + \eta_+^2(k))} - \ln \eta_+(k) - \gamma_E, \quad (30)$$

where $H(x)$ is the analytic function,

$$H(x) = \psi(x) + \frac{1}{2x} - \ln(x). \qquad (31)$$

Equating interior and exterior logarithmic derivatives as before, we find now

$$KF(KR) = \frac{\cos\delta_0 F_0^{+\prime}(R) - \sin\delta_0 G_0^{+\prime}(R)}{\cos\delta_0 F_0^+(R) - \sin\delta_0 G_0^+(R)}$$
$$= \frac{k\cot\delta_0 K_s \frac{1}{2a_B} - \frac{1}{R^2}}{k\cot\delta_0 K_s + \frac{1}{R} + \frac{1}{a_B}[h(\eta_+(k)) - \ln\frac{a_B}{R} + 2\gamma_E - 1]}.$$
$$(32)$$

Since $R \sim 1\,\mathrm{fm} \ll a_B \sim 50\,\mathrm{fm}$, eq. (32) can be written in the form

$$k\cot\delta_0 K_s + \frac{1}{a_B} \left[h(\eta_+(k)) - \ln \frac{a_B}{R} + 2\gamma_E - 1 \right]$$
$$\simeq -\frac{1}{a_0}. \qquad (33)$$

The scattering length a_C in the presence of the Coulomb interaction is conventionally defined as [25]

$$k\cot\delta_0 K_s + \frac{1}{a_B} h(\eta_+(k)) = -\frac{1}{a_C} + \ldots, \qquad (34)$$

so that we have the relation

$$-\frac{1}{a_0} = -\frac{1}{a_C} - \frac{1}{a_B} \left(\ln \frac{a_B}{R} + 1 - 2\gamma_E \right) \qquad (35)$$

between the experimental scattering length —a_C— and that which would exist in the absence of the Coulomb interaction —a_0.

As an aside, we note that a_0 is *not* itself an observable since the Coulomb interaction *cannot* be turned off. However, we can imagine a "gedanken scattering" in which there exists no Coulomb repulsion. In this case isotopic spin invariance requires the equality of the S-wave pp and nn scattering lengths —$a_0^{pp} = a_0^{nn}$— yielding the prediction

$$-\frac{1}{a_0^{nn}} = -\frac{1}{a_C^{pp}} - \alpha M_N \left(\ln \frac{1}{\alpha M_N R} + 1 - 2\gamma_E \right). \qquad (36)$$

While this is a model-dependent result, Jackson and Blatt have shown, by treating the interior Coulomb interaction perturbatively, that a version of this result with $1 - 2\gamma_E \rightarrow 0.824 - 2\gamma_E$ is approximately valid for a wide range of strong-interaction potentials [24] and the correction indicated in eq. (36) is essential in restoring agreement between the widely discrepant —$a_0^{nn} = -18.8\,\mathrm{fm}$ *vs.* $a_C^{pp} = -7.82\,\mathrm{fm}$— values obtained experimentally.

Returning to the problem at hand, the experimental scattering amplitude can then be written as

$$f_C^+(k) = \frac{e^{2i\sigma_0} K_s}{-\frac{1}{a_C} - \frac{1}{a_B} h(\eta_+(k)) - ikK_s}$$
$$= \frac{e^{2i\sigma_0} K_s}{-\frac{1}{a_C} - \frac{1}{a_B} H(i\eta_+(k))}, \qquad (37)$$

where $\sigma_0 = \arg \Gamma(1 - i\eta_+(k))$ is the S-wave Coulomb phase.

The above analysis is standard and can be found in many quantum mechanics texts [20]. In the next section we reanalyze the NN system using the ideas of effective field theory.

2.3 Effective field theory analysis

Identical results may be obtained using a parallel effective field theory (EFT) analysis and in many ways the derivation is clearer and more intuitive [26][2]. The basic idea here is that since we are only interested in interactions at very low energy, a scattering length description is quite adequate and it is unnecessary to identify a specific form for the potential —everything can be done in terms of observables. From eq. (11) we see that, at least for weak potentials, the scattering length has a natural representation in terms of the momentum space potential $\tilde{V}(\boldsymbol{p} = 0)$:

$$a_0 = \frac{m_r}{2\pi} \int \mathrm{d}^3 r V(r) = \frac{m_r}{2\pi} \tilde{V}(\boldsymbol{p} = 0) \qquad (38)$$

---

[2] Two interesting didactic introductions to EFT methods can be found in refs. [27] and [28].

Fig. 1. The multiple scattering series.

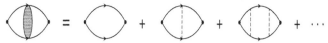

Fig. 2. The Coulomb-corrected bubble.

and it is thus natural to perform our analysis using a simple contact interation. First, consider the situation that we have two particles A, B interacting only via a *local* strong interaction, so that the effective Lagrangian can be written as

$$\mathcal{L} = \sum_{i=A}^{B} \Psi_i^\dagger \left(i\frac{\partial}{\partial t} + \frac{\boldsymbol{\nabla}^2}{2m_i} \right) \Psi_i - C_0 \Psi_A^\dagger \Psi_A \Psi_B^\dagger \Psi_B + \ldots \quad (39)$$

The T-matrix is then given in terms of the multiple scattering series shown in fig. 1

$$T_{fi}(k) = -\frac{2\pi}{m_r} f(k) = C_0 + C_0^2 G_0(k) + C_0^3 G_0^2(k) + \ldots$$
$$= \frac{C_0}{1 - C_0 G_0(k)}, \quad (40)$$

where $G_0(k)$ is the amplitude for particles A, B to travel from zero separation to zero separation —*i.e.*, the propagator $D_F(k; \boldsymbol{r}' = 0, \boldsymbol{r} = 0)$—

$$G_0(k) = \lim_{\boldsymbol{r}', \boldsymbol{r} \to 0} \int \frac{\mathrm{d}^3 s}{(2\pi)^3} \frac{e^{i\boldsymbol{s}\cdot\boldsymbol{r}'} e^{-i\boldsymbol{s}\cdot\boldsymbol{r}}}{\frac{k^2}{2m_r} - \frac{s^2}{2m_r} + i\epsilon}$$
$$= \int \frac{\mathrm{d}^3 s}{(2\pi)^3} \frac{2m_r}{k^2 - s^2 + i\epsilon} \quad (41)$$

(Equivalently $T_{fi}(k)$ satisfies a Lippman-Schwinger equation

$$T_{fi}(k) = C_0 + C_0 G_0(k) T_{fi}(k), \quad (42)$$

whose solution is given in eq. (40).)

The complication here is that the function $G_0(k)$ is divergent and must be defined via some sort of regularization scheme. There are a number of ways by which to accomplish this, but perhaps the simplest is to use a cutoff regularization with $k_{max} = \mu$, which simply eliminates the high-momentum components of the wave function completely. Then

$$G_0(k) = -\frac{m_r}{2\pi} \left(\frac{2\mu}{\pi} + ik \right) \quad (43)$$

(Other regularization schemes are similar. For example, one could subtract at an unphysical momentum point, as proposed by Gegelia [29]

$$G_0(k) = \int \frac{\mathrm{d}^3 s}{(2\pi)^3} \left(\frac{2m_r}{k^2 - s^2 + i\epsilon} + \frac{2m_r}{\Lambda^2 + s^2} \right) = -\frac{m_r}{2\pi}(\Lambda + ik), \quad (44)$$

which has been shown by Mehen and Stewart [30] to be equivalent to the power divergence subtraction scheme

proposed by Kaplan, Savage and Wise [26].) In any case, the would-be linear divergence is canceled by the introduction of a counterterm, which accounts for the omitted high-energy component of the theory and modifies C_0 to $C_0(\mu)$. (That $C_0(\mu)$ should be a function of the cutoff is clear because by varying the cutoff energy we are varying the amount of higher-energy physics which we are including in our effective description.) The scattering amplitude then becomes

$$f(k) = -\frac{m_r}{2\pi} \left(\frac{1}{\frac{1}{C_0(\mu)} - G_0(k)} \right) = \frac{1}{-\frac{2\pi}{m_r C_0(\mu)} - \frac{2\mu}{\pi} - ik}. \quad (45)$$

Comparing with eq. (2) we identify the scattering length as

$$-\frac{1}{a_0} = -\frac{2\pi}{m_r C_0(\mu)} - \frac{2\mu}{\pi}. \quad (46)$$

Of course, since a_0 is a physical observable, it must be *cutoff-independent* —the μ-dependence of $1/C_0(\mu)$ is precisely canceled by the cutoff dependence in the Green's function.

2.4 Coulomb effects in EFT

More interesting (and challenging) is the case where we restore the Coulomb interaction between the particles. The derivatives in eq. (39) then become covariant and the bubble sum is evaluated with static photon exchanges between each of the lines —each bubble is replaced by one involving a sum of zero, one, two, etc. Coulomb interactions, as shown in fig. 2.

The net result in the case of same charge scattering is the replacement of the free propagator by its Coulomb analog

$$G_0(k) \to G_C^+(k) = \lim_{\boldsymbol{r}', \boldsymbol{r} \to 0} \int \frac{\mathrm{d}^3 s}{(2\pi)^3} \frac{\psi_{\boldsymbol{s}}^+(\boldsymbol{r}') \psi_{\boldsymbol{s}}^{+*}(\boldsymbol{r})}{\frac{k^2}{2m_r} - \frac{s^2}{2m_r} + i\epsilon}$$
$$= \int \frac{\mathrm{d}^3 s}{(2\pi)^3} \frac{2m_r K_s}{k^2 - s^2 + i\epsilon}, \quad (47)$$

where

$$\psi_{\boldsymbol{s}}^+(\boldsymbol{r}) = C(\eta_+(s)) e^{i\sigma_0} e^{i\boldsymbol{s}\cdot\boldsymbol{r}} {}_1F_1(-i\eta_+(s), 1, isr - i\boldsymbol{s}\cdot\boldsymbol{r}) \quad (48)$$

is the outgoing Coulomb wave function for repulsive Coulomb scattering [31]. Also in the initial and final states the influence of static photon exchanges must be included to all orders, which produces the factor $K_s \exp(2i\sigma_0)$. Thus, the repulsive Coulomb scattering amplitude becomes

$$f_C^+(k) = -\frac{m_r}{2\pi} \frac{C_0 K_s \exp 2i\sigma_0}{1 - C_0 G_C^+(k)}. \quad (49)$$

The momentum integration in eq. (47) can be performed as before using cutoff regularization, yielding [32]

$$G_C^+(k) = -\frac{m_r}{2\pi}\left\{\frac{2\mu}{\pi} + \frac{1}{a_B}\left[H(i\eta_+(k)) - \ln\frac{\mu a_B}{\pi} - \zeta\right]\right\},\tag{50}$$

where $\zeta = \ln 2\pi - \gamma$. We have then

$$
\begin{aligned}
f_C^+(k) &= \frac{K_s e^{2i\sigma_0}}{-\frac{2\pi}{m_r C_0(\mu)} - \frac{2\mu}{\pi} - \frac{1}{a_B}[H(i\eta_+(k)) - \ln\frac{\mu a_B}{\pi} - \zeta]} \\
&= \frac{K_s e^{2i\sigma_0}}{-\frac{1}{a_0} - \frac{1}{a_B}\left[h(\eta_+(k)) - \ln\frac{\mu a_B}{\pi} - \zeta\right] - ikK_s}.
\end{aligned}\tag{51}
$$

Comparing with eq. (37) we identify the Coulomb scattering length as

$$-\frac{1}{a_C} = -\frac{1}{a_0} + \frac{1}{a_B}\left(\ln\frac{\mu a_B}{\pi} + \zeta\right)\tag{52}$$

which matches nicely with eq. (35) if a reasonable cutoff $\mu \sim m_\pi \sim 1/R$ is employed. The scattering amplitude then has the simple form

$$f_C^+(k) = \frac{K_s e^{2i\sigma_0}}{-\frac{1}{a_C} - \frac{1}{a_B}H(i\eta_+(k))}\tag{53}$$

in agreement with eq. (37).

The important lesson here is that at very low energy, where we can completely characterize the amplitude in terms of the scattering length, we see that only *two* parameters are required in order to completely describe NN scattering —the spin-singlet and triplet scattering lengths. The effective range parameters in these channels provide a way to estimate the size of possible corrections to this scattering length approximation. It is *not* necessary to make any assumptions about the detailed shape of the potential —we can write everything in terms of *observables*.

Our next goal then is to emulate this discussion in the case of the parity-*violating* NN potential, a task which we take up in the following section.

2.5 Parity-violating NN interaction: potential model description

Until recently the standard method by which to treat the parity-violating NN interaction was by use of potential theory. The basic idea is that in the same way in which the low-energy parity-conserving NN interaction can be described quite satisfactorily in terms of a simple light meson exchange picture, we can represent the parity-violating NN interaction in a parallel fashion, wherein one of the parity-conserving NNM vertices is replaced by its parity-violating analog —cf. fig. 3. This is the method pioneered by Blin-Stoyle and by Michel and then followed by DDH in their seminal 1980 paper. Of course, this approach is model dependent and generates a specific form for the potential, but in the days before effective field theory this was the standard way to proceed.

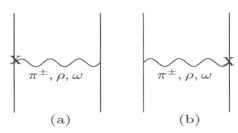

Fig. 3. Parity-violating NN potential generated by meson exchange. Here the symbol "X" indicates a parity-violating vertex.

More specifically, we represent the (parity-conserving) strong coupling of the nucleon to the light vector and pseudoscalar mesons via the effective Lagrangian

$$
\begin{aligned}
\mathcal{H}_{st} =\ &ig_{\pi NN}\bar{N}\gamma_5\tau\cdot\pi N + g_\rho\bar{N}\left(\gamma_\mu + i\frac{\chi_\rho}{2m_N}\sigma_{\mu\nu}k^\nu\right)\tau\cdot\rho^\mu N \\
&+ g_\omega\bar{N}\left(\gamma_\mu + i\frac{\chi_\omega}{2m_N}\sigma_{\mu\nu}k^\nu\right)\omega^\mu N,
\end{aligned}\tag{54}
$$

whose values are determined from strong-interaction studies. Typical —though not universally accepted [33]— values are $g_{\pi NN}^2/4\pi \simeq 13.5$ and $g_\rho^2/4\pi = \frac{1}{9}g_\omega^2/4\pi \simeq 0.67$ and, with the use of vector dominance to connect with the electromagnetic interaction, $\chi_\rho = \kappa_p - \kappa_n = 3.7$ and $\chi_\omega = \kappa_p + \kappa_n = -0.12$. For the parity-violating couplings we can write a general phenomenological interaction of the form [17]

$$
\begin{aligned}
\mathcal{H}_{wk} =\ &i\frac{f_\pi^1}{\sqrt{2}}\bar{N}(\tau\times\pi)_z N \\
&+ \bar{N}\left(h_\rho^0\tau\cdot\rho^\mu + h_\rho^1\rho_z^\mu + \frac{h_\rho^2}{2\sqrt{6}}(3\tau_z\rho_z^\mu - \tau\cdot\rho^\mu)\right)\gamma_\mu\gamma_5 N \\
&+ \bar{N}\left(h_\omega^0\omega^\mu + h_\omega^1\tau_z\omega^\mu\right)\gamma_\mu\gamma_5 N - h_\rho^{\prime 1}\bar{N}(\tau\times\rho^\mu)_z\frac{\sigma_{\mu\nu}k^\nu}{2m_N}\gamma_5 N,
\end{aligned}\tag{55}
$$

where here we have used the stricture from Barton's theorem that any CP-conserving parity-violating coupling to neutral pseudoscalar mesons such as π^0, η^0 must vanish [34]. We see then that there exist, in this model, *seven* unknown weak couplings f_π^1, $h_\rho^{(0)}$, $h_\rho^{(1)}$, $h_\rho^{(2)}$, $h_\omega^{(0)}$, $h_\omega^{(1)}$, $h_\rho^{(1)'}$. However, quark model calculations suggest that $h_\rho^{(1)'}$ is quite small [35], so this term is generally omitted, leaving parity-violating observables described in terms of just six phenomenological constants —f_π, $h_\rho^{(0)}$, $h_\rho^{(1)}$, $h_\rho^{(2)}$, $h_\omega^{(0)}$, $h_\omega^{(1)}$[3]. In their paper DDH attempted to evaluate these basic PV couplings using basic quark-model and $SU(6)$-symmetry techniques, but they encountered significant theoretical challenges and uncertainties. For this reason their results were presented in terms of an allowable range for each, accompanied by a "best value" representing their reasonable guess for each coupling. These

---

[3] Another way to view the neglect of the $h_\rho^{\prime 1}$ coupling is that it represents simply a short-range correction to the size of the charged-pion exchange coupling.

Table 1. Weak NNM couplings as calculated in refs. [17,36,37]. All numbers are quoted in units of the "sum rule" value $g_\pi = 3.8 \cdot 10^{-8}$.

Coupling	DDH [17] Reasonable range	DDH [17] "Best" value	DZ [36]	FCDH [37]
f_π	$0 \to 30$	$+12$	$+3$	$+7$
h_ρ^0	$30 \to -81$	-30	-22	-10
h_ρ^1	$-1 \to 0$	-0.5	$+1$	-1
h_ρ^2	$-20 \to -29$	-25	-18	-18
h_ω^0	$15 \to -27$	-5	-10	-13
h_ω^1	$-5 \to -2$	-3	-6	-6

ranges and "best values" are listed in table 1, together with predictions generated by subsequent groups [36,37]. (This list is not comprehensive, merely representative, and many other estimates have been provided. For example, Kaiser and Meissner utilized a chiral soliton approach to calculate these numbers [38], while Hwang and Wen employed the method of QCD sum rules to yield values for the DDH couplings [39].)"

Before making contact with experimental results, however, it is necessary to convert the NNM couplings generated above into an effective parity-violating NN potential. Inserting the strong and weak couplings, defined above into the meson exchange diagrams shown in fig. 1 and transforming to coordinate space, one finds the DDH parity-violating NN potential

$$V_{DDH}^{PV}(r) = i \frac{f_\pi^1 g_{\pi NN}}{\sqrt{2}} \left(\frac{\tau_1 \times \tau_2}{2} \right)_z (\sigma_1 + \sigma_2) \cdot \left[\frac{p_1 - p_2}{2m_N}, w_\pi(r) \right]$$

$$- g_\rho \left(h_\rho^0 \tau_1 \cdot \tau_2 + h_\rho^1 \left(\frac{\tau_1 + \tau_2}{2} \right)_z + h_\rho^2 \frac{(3\tau_1^z \tau_2^z - \tau_1 \cdot \tau_2)}{2\sqrt{6}} \right)$$

$$\times \left((\sigma_1 - \sigma_2) \cdot \left\{ \frac{p_1 - p_2}{2m_N}, w_\rho(r) \right\} \right.$$

$$+ i(1 + \chi_V) \sigma_1 \times \sigma_2 \cdot \left[\frac{p_1 - p_2}{2m_N}, w_\rho(r) \right] \right)$$

$$- g_\omega \left(h_\omega^0 + h_\omega^1 \left(\frac{\tau_1 + \tau_2}{2} \right)_z \right)$$

$$\times \left((\sigma_1 - \sigma_2) \cdot \left\{ \frac{p_1 - p_2}{2m_N}, w_\omega(r) \right\} \right.$$

$$+ i(1 + \chi_S) \sigma_1 \times \sigma_2 \cdot \left[\frac{p_1 - p_2}{2m_N}, w_\omega(r) \right] \right) - \left(g_\omega h_\omega^1 - g_\rho h_\rho^1 \right)$$

$$\times \left(\frac{\tau_1 - \tau_2}{2} \right)_z (\sigma_1 + \sigma_2) \cdot \left\{ \frac{p_1 - p_2}{2m_N}, w_\rho(r) \right\}$$

$$- g_\rho h_\rho^{1'} i \left(\frac{\tau_1 \times \tau_2}{2} \right)_z (\sigma_1 + \sigma_2) \cdot \left[\frac{p_1 - p_2}{2m_N}, w_\rho(r) \right], \quad (56)$$

where $w_i(r) = \exp(-m_i r)/4\pi r$ is the usual Yukawa form, $r = |x_1 - x_2|$ is the separation between the two nucleons, and $p_i = -i\nabla_i$. We observe from eq. (56) that the unknown weak couplings f_π^1, h_V^I always occur multiplied by their strong-interaction counterparts $g_{\pi NN}, g_V, g_V \chi_V$ so that the lack of precise knowledge of these strong cou-

plings alluded to above does not really damage the use of the DDH potential for phenomenological purposes.

It is useful to note at this point that the DDH model is the parity-violating analog of the conventional potential approach to scattering and postulates a *complete* form of the effective parity-violating NN potential —both magnitude and shape— so that in principle even high-energy observables are predicted. (In reality, of course, high-energy forms should include meson exchanges from heavier systems such as the axial mesons.) It is also important to point out that each of the vector-meson–mediated pieces of the potential consists of *both* a convective (anticommutator) *and* magnetic (commutator) component, with the relative strength of these two couplings determined from vector dominance in terms of the anomalous magnetic moments of the nucleons, as outlined above.

Essentially *all* experimental results involving hadronic parity violation have been analyzed using $V_{DDH}^{PV}(r)$ for the past quarter century. There have been a number of previous reviews of this field, beginning with the 1985 *Annual Reviews of Nuclear and Particle Science* article by Adelberger and Haxton [40], continuing with the 1995 review by Holstein and Haeberli appearing in the book *Symmetries and Fundamental Interactions in Nuclei* [41], and in 2006 the field was again surveyed by Page and Ramsey-Musolf in *Annual Reviews* [42]. Because each of these papers *comprehensively* examines the experimental situation in a fashion far deeper than possible in the present article, we defer to them for details of the various experiments and merely report here the conclusions, which are that, despite half a century of experimental and theoretical work, at present there appear to exist significant discrepancies between the values extracted for the various DDH couplings from different experiments.

The problem can be seen in a number of ways, but perhaps the most straightforward is to note that analysis of experiments on the asymmetry in longitudinally polarized $p\alpha$ scattering [43] and the photon asymmetry in the decay of the polarized first excited state of ^{19}F [44] are consistent with each other and (within errors) with values about half of the best guess DDH numbers. The analysis depends predominantly on the long-range pion coupling f_π^1 and on the effective isoscalar vector meson coupling $h_\rho^0 + 0.7h_\omega^0$ and is often presented in terms of a two-dimensional plot —cf. fig. 4. However, at least four

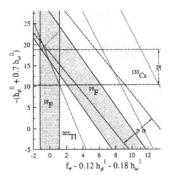

Fig. 4. Experimental limits on weak couplings.

experiments seeking the circular polarization of photons emitted in the decay of the 1.081 MeV excited state of ^{18}F have failed to see any signal [45], which seems to indicate that the pion coupling f_π^1 is considerably *smaller* than its DDH best guess value. One might be tempted to attribute this inconsistency to nuclear uncertainties, but the theoretical analysis of this mode is buttressed by comparison with the two-body contributions to the analog beta decay of ^{18}Ne [46] as well as by a very recent cold neutron experiment which measured the triton asymmetry in the reaction ^{6}Li$(n, \alpha)^3$H [47]. An additional issue is that recent measurements of the anapole moment of ^{133}Cs from atomic parity violation experiments [48] appear to be consistent with a size for f_π^1 in agreement with the DDH "best value". This situation is summarized in fig. 4, which clearly indicates difficulties with the present DDH analysis of the PV NN interaction. These discrepancies possibly suggest a problem with the underlying model-dependent theoretical framework itself, and it is for this reason that a new approach, based on effective field theory, has been developed. This technique is discussed below.

2.6 Parity-violating NN interaction: EFT description

As described above, there presently exist inconsistencies within the DDH analysis of $\Delta S = 0$ hadronic parity-violating experiments. The origin of this problem is unclear, but could certainly be associated with the fact that the extraction of the basic weak couplings requires knowledge of the spatial average of the associated weak parity-violating potential weighted by imperfectly known nuclear wave functions. For this reason and for basic understanding of such processes, the analysis of such experiments has been recently reformulated in terms of an effective field-theoretic parity-violating NN potential, which puts the analysis of these systems into a more rigorous model-independent form. The basic idea here is that there are a number of scales at play in the manifestations of the hadronic weak interaction. There is the momentum transfer Q which is generally much smaller than the chiral scale $\Lambda_\chi \simeq 4\pi F_\pi$ but can be smaller than or comparable to the inverse nucleon size $1/R \sim m_\pi$. First, suppose that $Q \ll m_\pi \ll \Lambda_\chi$. In order to formulate the EFT discussion in parallel to that used in the analysis

of the parity-conserving interaction, we begin by writing down the lowest-order (one-derivative) short-range (SR) form of the PV NN potential $V^{PV}_{1,SR}(\mathbf{r})$, as given by Zhu *et al.* [18][4],

$$
\begin{aligned}
V^{PV}_{1,SR}(\mathbf{r}) = \frac{2}{\Lambda_\chi^3} &\left\{ \left[C_1 + C_2 \frac{\tau_1^z + \tau_2^z}{2} \right] (\boldsymbol{\sigma}_1 - \boldsymbol{\sigma}_2) \cdot \{-i\boldsymbol{\nabla}, f_m(r)\} \right. \\
&+ \left[\tilde{C}_1 + \tilde{C}_2 \frac{\tau_1^z + \tau_2^z}{2} \right] i(\boldsymbol{\sigma}_1 \times \boldsymbol{\sigma}_2) \cdot [-i\boldsymbol{\nabla}, f_m(r)] \\
&+ [C_2 - C_4] \frac{\tau_1^z - \tau_2^z}{2} (\boldsymbol{\sigma}_1 + \boldsymbol{\sigma}_2) \cdot \{-i\boldsymbol{\nabla}, f_m(r)\} \\
&+ \left[C_3 \boldsymbol{\tau}_1 \cdot \boldsymbol{\tau}_2 + C_4 \frac{\tau_1^z + \tau_2^z}{2} + \mathcal{I}_{ab} C_5 \tau_1^a \tau_2^b \right] \\
&\quad \times (\boldsymbol{\sigma}_1 - \boldsymbol{\sigma}_2) \cdot \{-i\boldsymbol{\nabla}, f_m(r)\} \\
&+ \left[\tilde{C}_3 \boldsymbol{\tau}_1 \cdot \boldsymbol{\tau}_2 + \tilde{C}_4 \frac{\tau_1^z + \tau_2^z}{2} + \mathcal{I}_{ab} \tilde{C}_5 \tau_1^a \tau_2^b \right] \\
&\quad \times i(\boldsymbol{\sigma}_1 \times \boldsymbol{\sigma}_2) \cdot [-i\boldsymbol{\nabla}, f_m(r)] \\
&\left. + \tilde{C}_6 i(\boldsymbol{\tau}_1 \times \boldsymbol{\tau}_2)_z (\boldsymbol{\sigma}_1 + \boldsymbol{\sigma}_2) \cdot [-i\boldsymbol{\nabla}, f_m(r)] \right\}, \quad (57)
\end{aligned}
$$

which is the PV analog of the contact PC interaction eq. (39), the derivative form in eq. (57) being required by the stricture of parity violation. Here

$$
\mathcal{I}^{ab} = \begin{pmatrix} 1 & 0 & 0 \\ 0 & 1 & 0 \\ 0 & 0 & -2 \end{pmatrix}, \quad (58)
$$

and $f_m(\mathbf{r})$ is a function which

i) is strongly peaked, with width $\sim 1/m$ about $r = 0$, and

ii) approaches $\delta^{(3)}(\mathbf{r})$ in the zero width —$m \to \infty$— limit.

A convenient (though not unique) form, for example, is the Yukawa-like function

$$
f_m(r) = \frac{m^2}{4\pi r} \exp(-mr), \quad (59)
$$

where m is a mass chosen to reproduce the appropriate short-range effects. (Actually, for the purpose of carrying out actual calculations, one could just as easily use the momentum space form of V^{PV}_{SR}, thereby avoiding the use of $f_m(\mathbf{r})$ altogether.)

The matching of the DDH model to the coefficients C_i, $\tilde{C}_i$ in eq. (57) can be done by writing the Yukawa functions $w_i(r)$ in terms of their Fourier transforms

$$
w_i(r) = \int \frac{\mathrm{d}^3 Q}{(2\pi)^3} \frac{e^{i\mathbf{Q} \cdot \mathbf{r}}}{m_i^2 + \mathbf{Q}^2}. \quad (60)
$$

---

[4] Note that below, as suggested by Liu [49], we have used the symbol $\tilde{C}_6$ rather than C_6 as used by Zhu *et al.* [18] since it multiplies a commutator rather than an anticommutator.

Working in the limit in which $\boldsymbol{Q}^2 \ll mi^2$ with $i = \pi, \rho, \omega$ we can make the replacement $m_i^2 + \boldsymbol{Q}^2 \longrightarrow m_i^2$, whereby the Yukawa function is replaced by a delta function,

$$w_i(r) \longrightarrow \frac{1}{m_i^2} \delta^3(r).$$

In the Zhu *et al.* formalism this delta function is represented by $f_m(r)$. We observe then that the same set of spin-space and isospin structures appear in both V_{eff}^{PV} and the vector-meson exchange terms in V_{DDH}^{PV}, though the relationship between the various coefficients in V_{eff}^{PV} is more general. In particular, the DDH model is tantamount to assuming

$$\frac{\tilde{C}_1}{C_1} = \frac{\tilde{C}_2}{C_2} = 1 + \chi_\omega \simeq 0.88, \tag{61}$$

$$\frac{\tilde{C}_3}{C_3} = \frac{\tilde{C}_4}{C_4} = \frac{\tilde{C}_5}{C_5} = 1 + \chi_\rho \simeq 4.7, \tag{62}$$

and taking $m \sim m_\rho$, m_ω for $C_1, \dots, C_5$ but $m \sim m_\pi$ for $\tilde{C}_6$, assumptions which may not be physically realistic. Nevertheless, if this ansatz is posited, the EFT and DDH results coincide provided the identifications

$$C_1^{DDH} = -\frac{\Lambda_\chi^3}{2m_N m_\omega^2} g_\omega h_\omega^0 \xrightarrow{\text{best guess}} 2.3 \times 10^{-6},$$

$$C_2^{DDH} = -\frac{\Lambda_\chi^3}{2m_N m_\omega^2} g_\omega h_\omega^1 \xrightarrow{\text{best guess}} 1.4 \times 10^{-6},$$

$$C_3^{DDH} = -\frac{\Lambda_\chi^3}{2m_N m_\rho^2} g_\rho h_\rho^0 \xrightarrow{\text{best guess}} 4.6 \times 10^{-6},$$

$$C_4^{DDH} = -\frac{\Lambda_\chi^3}{2m_N m_\rho^2} g_\rho h_\rho^1 \xrightarrow{\text{best guess}} 0.1 \times 10^{-6},$$

$$C_5^{DDH} = \frac{\Lambda_\chi^3}{4\sqrt{6}m_N m_\rho^2} g_\rho h_\rho^2 \xrightarrow{\text{best guess}} -0.8 \times 10^{-6},$$

$$\tilde{C}_6^{DDH} \simeq \tilde{C}_6^\pi = \frac{\Lambda_\chi^3}{2\sqrt{2}m_N m_\pi^2} g_{\pi NN} f_\pi^1 \xrightarrow{\text{best guess}} 180 \times 10^{-6}, \tag{63}$$

are made [18] and only the S-P mixing terms in the DDH form are retained. (Note that for use below we have quoted the "best value" numbers for these parameters.)

This form of the effective theory is generally termed the "pionless" picture because, since $Q \ll m_\pi$, the pion does not appear as an explicit degree of freedom.

Of course, the "pionless" approximation breaks down at energies of order $m_\pi^2/M_N \sim 20\,\text{MeV}$ and must be replaced by a somewhat more complex theory which *does* contain an explicit pion. In this "pionful" theory, which should work until energies of order the pion mass, we have $\boldsymbol{Q}^2 \simeq m_\pi^2$, but we still have $Q^2 \ll m_\rho^2 m_\omega^2$ so that

the matching of the DDH picture to the effective theory given above is unchanged for the coefficients C_i, $\tilde{C}_i$ $i = 1, 2, \dots, 5$. However, in the case of $\tilde{C}_6$, which is associated with pion exchange, the replacement of this piece by an effective short-range interaction is no longer justified. Instead the last line of eq. (57) —i.e. the term involving $\tilde{C}_6$ must be removed and replaced by four additional types of terms:

i) a long-range one-pion exchange potential, which is two orders *lower* in chiral counting than the corresponding short-range (vector-meson exchange) terms:

$$V_{-1,LR}(r) = \frac{1}{\Lambda_\chi^3} \tilde{C}_6^\pi i(\boldsymbol{\tau}_1 \times \boldsymbol{\tau}_2)_z (\boldsymbol{\sigma}_1 + \boldsymbol{\sigma}_2) \cdot [-i\boldsymbol{\nabla}, f_\pi(r)], \tag{64}$$

where $\tilde{C}_6^\pi$ is defined in eq. (63).

ii) a medium-range interaction which arises from the effects of two-pion exchange

$$V_{1,MR}(r) = \frac{\tilde{C}_2^{2\pi}}{\Lambda_\chi^3} \Bigg\{ (\tau_1 + \tau_2)_z i(\boldsymbol{\sigma}_1 \times \boldsymbol{\sigma}_2) \cdot \boldsymbol{y}_{2\pi}^L(r)$$
$$- \frac{3}{4k}(\tau_1 \times \tau_2)_z (\boldsymbol{\sigma}_1 + \boldsymbol{\sigma}_2)$$
$$\times \left[\left(1 - \frac{1}{3g_A^2}\right) \boldsymbol{y}_{2\pi}^L(r) - \frac{1}{3}\boldsymbol{y}_{2\pi}^H(r) \right] \Bigg\}, \tag{65}$$

where

$$\tilde{C}_2^{2\pi} = -4\sqrt{2}\pi g_A^3 f_\pi^1 \tag{66}$$

and the functions $\boldsymbol{y}_{2\pi}^{H,R}(r)$ are defined via

$$\boldsymbol{y}_{2\pi}^{H,L}(r) = [-i\boldsymbol{\nabla}, H, L(r)] \tag{67}$$

with $H, L(r)$ being the Fourier transform of the functions

$$L(\boldsymbol{q}) = \frac{\sqrt{4m_\pi^2 + \boldsymbol{q}^2}}{|\boldsymbol{q}|} \log\left(\frac{\sqrt{4m_\pi^2 + \boldsymbol{q}|^2} + |\boldsymbol{q}|}{2m_\pi} \right),$$

$$H(\boldsymbol{q}) = \frac{4m_\pi^2}{4m_\pi^2 + \boldsymbol{q}^2} L(\boldsymbol{q}), \tag{68}$$

respectively. (Note that these terms include only the *nonanalytic* pieces of the full two-pion exchange amplitude, since it is only these pieces which yield medium-range effects. Any analytic component of the two-pion exchange amplitude generates a short-distance contribution, which is subsumed into the phenomenological coefficients of the contact terms already written down.)

iii) a long-range component generated from one-loop corrections to the leading vertices, which is two orders higher in the counting than the leading pion-exchange potential:

$$
V_{1,LR}(\boldsymbol{p}_1, \boldsymbol{p}'_1, \boldsymbol{p}_2, \boldsymbol{p}'_2) = \frac{g_A h^1_\pi}{\Lambda_\chi F^2_\pi} \frac{1}{2} (\boldsymbol{\tau}_1 \times \boldsymbol{\tau}_2)_z
$$
$$
\times \left[\frac{\boldsymbol{\sigma}_1 \cdot \boldsymbol{p}'_1 \times \boldsymbol{p}_1 \boldsymbol{\sigma}_2 \cdot \boldsymbol{q}}{q^2 + m^2_\pi} + (1 \leftrightarrow 2) \right]
$$
$$
+ i \frac{g_A f^1_\pi}{\sqrt{2} m^2_N F_\pi} \frac{1}{2} (\boldsymbol{\tau}_1 \times \boldsymbol{\tau}_2)_z \frac{1}{q^2 + m^2_\pi}
$$
$$
\times \left\{ \frac{1}{4} [(\boldsymbol{p}^2_1 - \boldsymbol{p}'^2_1) \boldsymbol{\sigma}_1 \cdot (\boldsymbol{p}'_1 + \boldsymbol{p}_1) - (1 \leftrightarrow 2)] \right.
$$
$$
- \frac{1}{8} [(\boldsymbol{p}^2_1 + \boldsymbol{p}'^2_1) \boldsymbol{\sigma}_1 \cdot \boldsymbol{q} + (1 \leftrightarrow 2)]
$$
$$
\left. + \frac{1}{4} [\boldsymbol{\sigma} \cdot \boldsymbol{p}^1_1 \boldsymbol{q} \cdot \boldsymbol{p}_1 + \boldsymbol{\sigma} \cdot \boldsymbol{p}_1 \boldsymbol{q} \cdot \boldsymbol{p}'_1 + (1 \leftrightarrow 2)] \right\}, \quad (69)
$$

where $\boldsymbol{q}_i = \boldsymbol{p}'_i - \boldsymbol{p}_i$.

iv) a PV "Kroll-Ruderman"-like $NN\pi\gamma$ coupling $\tilde{C}_\pi$ that leads to a new independent current operator:

$$
\boldsymbol{J}(\boldsymbol{x}_1, \boldsymbol{x}_2) = \frac{\sqrt{2} g_A \tilde{C}_\pi m^2_\pi}{\Lambda^2_\chi F_\pi} e^{-i\boldsymbol{q} \cdot \boldsymbol{x}_1} \tau^+_1 \tau^-_2 \boldsymbol{\sigma}_1 \times \boldsymbol{q} \boldsymbol{\sigma}_2 \cdot \hat{r} H_\pi(r)
$$
$$
+ (1 \leftrightarrow 2), \quad (70)
$$

where

$$
H_\pi(r) = \frac{e^{-m_\pi r}}{m_\pi r} \left(1 + \frac{1}{m_\pi r} \right). \quad (71)
$$

In this pionful theory the parameter m which characterizes nonpionic pieces of the potential (which is of order m_π in the pionless theory) should presumably assume a value of order $m \sim m_\rho \sim m_\omega$, since the pionic degrees of freedom are included explicitly in the potentials i), ii), iii), iv) described above.

It is interesting to note here that the "best guess" parameter $\tilde{C}^\pi_6$ is at least an order of magnitude larger than any of its short-distance (vector-meson–dominated) counterparts or than the medium-range "best guess" values, as might be suspected from its lower order in the chiral counting scheme.

3 Danilov parameters

The discussion in the previous section might make it appear that the low-energy analysis of the PV NN interaction must involve the determination of *ten* parameters[5] —a daunting task indeed. However, this assumption is misleading. *In fact, it is easy to see that at the very lowest energies there can be only five phenomenological constants involved.* This is because at threshold energies, we can neglect all but S-P-wave mixing, in which case there exist only *five* independent phenomenological amplitudes:

i) $d_t(k)$ representing $^3S_1$-$^1P_1$ mixing with $\Delta I = 0$;

---

[5] There appear to exist *eleven* terms —C_i, $\tilde{C}_i$, $i = 1, \ldots, 5$ plus $\tilde{C}_6$. However, $\tilde{C}_2$ and $\tilde{C}_4$ appear only in the combination $\tilde{C}_2 + \tilde{C}_4$.

ii) $d^{0,1,2}_s(k)$ representing $^1S_0$-$^3P_0$ mixing with $\Delta I = 0, 1, 2$ respectively;

iii) $c_t(k)$ representing $^3S_1$-$^3P_1$ mixing with $\Delta I = 1$.

These *five* independent transition amplitudes are the PV analogs of the *two* (singlet and triplet) S-wave amplitudes $m_s(k)$, $m_t(k)$ involved in the PC case.

Following Danilov [19], the low-energy parity-violating scattering matrix in the presence of parity violation can be written then as[6]

$$
\mathcal{M}_{PV}(\boldsymbol{k}', \boldsymbol{k}) = c_t(k)(\boldsymbol{\sigma}_1 + \boldsymbol{\sigma}_2) \cdot (\boldsymbol{k}' + \boldsymbol{k}) \frac{1}{2} (\boldsymbol{\tau}_1 - \boldsymbol{\tau}_2)_z
$$
$$
+ (\boldsymbol{\sigma}_1 - \boldsymbol{\sigma}_2) \cdot (\boldsymbol{k}' + \boldsymbol{k})
$$
$$
\times \left(P_0 d^0_s(k) + \frac{1}{2} (\tau_1 + \tau_2)_z d^1_s(k) + \frac{3\tau_{1z}\tau_{2z} - \boldsymbol{\tau}_1 \cdot \boldsymbol{\tau}_2}{2\sqrt{6}} d^2_s(k) \right)
$$
$$
+ d_t(k)(\boldsymbol{\sigma}_1 - \boldsymbol{\sigma}_2) \cdot (\boldsymbol{k}' + \boldsymbol{k}) P_1. \quad (72)
$$

Note that under spatial inversion —$\boldsymbol{\sigma} \to \boldsymbol{\sigma}$, $\boldsymbol{k}, \boldsymbol{k}' \to -\boldsymbol{k}, -\boldsymbol{k}'$— each of these pieces is P-odd, while under time reversal —$\boldsymbol{\sigma} \to -\boldsymbol{\sigma}$, $\boldsymbol{k}, \boldsymbol{k}' \to -\boldsymbol{k}', -\boldsymbol{k}$— each term is T-even. At very low energies the coefficients in the T-matrix become real and we can define [22]

$$
\lim_{k \to 0} c_t(k), d_s(k), d_t(k) \equiv \rho_t a_t, \lambda^i_s a_s, \lambda_t a_t. \quad (73)
$$

The motivation for inclusion of the S-wave scattering lengths a_t, a_s will be described presently. The five real numbers (Danilov parameters) ρ_t, λ^i_s, λ_t then completely characterize the lowest-energy parity-violating interaction and can in principle be determined experimentally, as we shall discuss below[7]. Alternatively, instead of a total isotopic spin representation, we can write things in terms of the equivalent notation

$$
\lambda^{pp}_s = \lambda^0_s + \lambda^1_s + \frac{1}{\sqrt{6}} \lambda^2_s,
$$
$$
\lambda^{np}_s = \lambda^0_s - \frac{2}{\sqrt{6}} \lambda^2_s,
$$
$$
\lambda^{nn}_s = \lambda^0_s - \lambda^1_s + \frac{1}{\sqrt{6}} \lambda^2_s. \quad (74)
$$

In Born approximation we can represent this interaction in terms of a simple effective NN potential. Integrating by parts, we have

$$
\int \mathrm{d}^3 r' \frac{e^{ik|\boldsymbol{r} - \boldsymbol{r}'|}}{|\boldsymbol{r} - \boldsymbol{r}'|} \{ -i\boldsymbol{\nabla}, \delta^3(\boldsymbol{r}') \} e^{i\boldsymbol{k} \cdot \boldsymbol{r}'} = (-i\boldsymbol{\nabla} + \boldsymbol{k}) \frac{e^{ikr}}{r} \quad (75)
$$

which represents the parity-violating contribution to the scattering wave function in terms of an S-wave admixture to the scattering P-wave state —$\sim \boldsymbol{\sigma} \cdot \boldsymbol{k} e^{ikr}/r$—

---

[6] An alternative low-energy form based on the Bethe-Goldstone equation has been given by Desplanques and Missimer [50].

[7] Note that there exists no singlet analog to the spin-triplet constant c_t since the combination $\boldsymbol{\sigma}_1 + \boldsymbol{\sigma}_2$ is proportional to the total spin operator and vanishes when operating on a spin-singlet state.

plus a P-wave admixture to the scattering S-state —$\sim -i\boldsymbol{\sigma} \cdot \boldsymbol{\nabla} e^{ikr}/r$. We observe then that the parity-violating component of the scattering wave function can be described via the effective potential

$$
U(\boldsymbol{r}) = \frac{4\pi}{M}\Big[\lambda_t a_t (\boldsymbol{\sigma}_1 - \boldsymbol{\sigma}_2) \cdot \{-i\boldsymbol{\nabla}, \delta^3(\boldsymbol{r})\}P_1
$$
$$
+\rho_t a_t (\boldsymbol{\sigma}_1 + \boldsymbol{\sigma}_2) \cdot \{-i\boldsymbol{\nabla}, \delta^3(\boldsymbol{r})\}\frac{1}{2}(\tau_1 - \tau_2)_z
$$
$$
+(\boldsymbol{\sigma}_1 - \boldsymbol{\sigma}_2) \cdot \{-i\boldsymbol{\nabla}, \delta^3(\boldsymbol{r})\}a_s
$$
$$
\times\left(P_0\lambda_s^0 + \frac{1}{2}(\tau_1 + \tau_2)_z\lambda_s^1 + \frac{3\tau_{1z}\tau_{2z} - \boldsymbol{\tau}_1\cdot\boldsymbol{\tau}_2}{2\sqrt{6}}\lambda_s^2\right)\Big],
$$
$$(76)$$

which is the PV analog of eq. (17).

Before application of this effective potential we must worry about the stricture of unitarity, which we have seen can be enforced in effective field theory language by using a Lippman-Schwinger solution. However, things become more interesting in the case of the parity-violating transitions, for which the requirement of unitarity reads, *e.g.*, for the case of scattering in the $^3S_1$-$^1P_1$ channel

$$
\mathrm{Im}\, d_t(k) = k[m_t^*(k)d_t(k) + d_t^*(k)m_p(k)], \qquad (77)
$$

where $m_p(k)$ is the $^1P_1$ analog of $m_t(k)$. Equation (77) is satisfied by the solution

$$
d_t(k) = |d_t(k)|e^{i(\delta_{3S_1}(k)+\delta_{1P_1}(k))} \qquad (78)
$$

i.e., the phase of the parity-violating transition amplitude should be the sum of the strong-interaction phases in the incoming and outgoing channels [51]. At very low energies we can neglect the P-wave phase and can write, following Danilov, the (approximately) unitarized forms

$$
c_t(k) \simeq \rho_t m_t(k), \quad d_s^i(k) \simeq \lambda_s^i m_s(k), \quad d_t(k) \simeq \lambda_t m_t(k).
$$
$$(79)$$

Since at threshold $m_t(k), m_s(k) \to a_t, a_s$, the threshold values of the parity-violating amplitudes become

$$
c_t(0) = \rho_t a_t, \qquad d_s^i(0) \simeq \lambda_s^i a_s, \qquad d_t(0) \simeq \lambda_t a_t \quad (80)
$$

and it is for this reason that the empirical S-P mixing parameters are defined by multiplying Danilov parameters by the relevant S-wave scattering lengths.

This result is also easily seen in the language of EFT, wherein the full transition matrix must include the weak amplitude to lowest order accompanied by rescattering in both incoming and outgoing channels to all orders in the strong interaction. If we represent the lowest-order weak contact interaction as

$$
T_{0tp}(k) = D_{0tp}(\mu)(\boldsymbol{\sigma}_1 - \boldsymbol{\sigma}_2) \cdot (\boldsymbol{k} + \boldsymbol{k}'), \qquad (81)
$$

then the full amplitude is given by

$$
T_{tp}(k) = \frac{D_{0tp}(\mu)}{(1 - C_{0t}(\mu)G_0(k))(1 - C_{0p}(\mu)G_1(k))}
$$
$$
\times(\boldsymbol{\sigma}_1 - \boldsymbol{\sigma}_2) \cdot (\boldsymbol{k} + \boldsymbol{k}'), \qquad (82)
$$

where we have introduced a lowest-order contact term C_{0p} which describes the $^1P_1$-wave nn interaction. Since the phase of the combination $1 - C_0(\mu)G_0(k)$ is simply the negative of the strong-interaction phase the unitarity stricture is clear, and we can define the physical transition amplitude A_{tp} via

$$
A_{tp} \equiv (1 + ika_t)(1 + ik^3 a_p)
$$
$$
\times \frac{D_{0tp}(\mu)}{(1 - C_{0t}(\mu)G_0(k))(1 - C_{0p}(\mu)G_1(k))}. \qquad (83)
$$

Making the identification $\lambda_t = -\frac{m_N}{4\pi}A_{tp}$ and noting that

$$
\frac{1}{1 + ika_t} = \cos\delta_t(k)e^{i\delta_t(k)}
$$

the Danilov parameter λ_t is seen to be identical to the R-matrix element defined by Miller and Driscoll [51].

The "mystery" of how *ten* contact terms —$C_i, \tilde{C}_i$— can be related to only *five* Danilov parameters can be solved by noting that the matrix elements of the commutator and anticommutator are identical in the zero-range (contact interaction) approximation —ZRA— in which $m \to \infty$[8]

$$
\lim_{m\to\infty}\langle P|[-i\boldsymbol{\nabla}, f_m(r)]|S\rangle = \lim_{m\to\infty}\langle P|\{-i\boldsymbol{\nabla}, f_m(r)\}|S\rangle.
$$
$$(84)$$

In this limit the contribution of the various operators characterized by $C_i\tilde{C}_i$ to observables can only occur in five different combinations, which may be found by use of the identity

$$
\boldsymbol{\sigma}_1 - \boldsymbol{\sigma}_2 = -\frac{i}{2}(\boldsymbol{\sigma}_1 \times \boldsymbol{\sigma}_2)(1 + \boldsymbol{\sigma}_1 \cdot \boldsymbol{\sigma}_2). \qquad (85)
$$

Thus for the $^1S_0$-$^3P_0$ parameters $d_s^{0,1,2}(k)$, we have $\boldsymbol{\sigma}_1 - \boldsymbol{\sigma}_2 = i\boldsymbol{\sigma}_1 \times \boldsymbol{\sigma}_2$ and we find that C_i and $\tilde{C}_i$ appear in the combination $C_i + \tilde{C}_i$ —*i.e.*, in the ZRA, the dependence in the different channels upon the EFT parameters $C_i, \tilde{C}_i$ must be

i) $^1S_0 \to {}^3P_0$ pp: $C_1 + C_2 + C_3 + C_4 - 2C_5 + (C_i \to \tilde{C}_i)$,
ii) $^1S_0 \to {}^3P_0$ nn: $C_1 - C_2 + C_3 - C_4 - 2C_5 + (C_i \to \tilde{C}_i)$,
iii) $^1S_0 \to {}^3P_0$ pn: $C_1 + C_3 + 4C_5 + (C_i \to \tilde{C}_i)$.

On the other hand, in the case of the $^3S_1$ parameter d_t we have $\boldsymbol{\sigma}_1 - \boldsymbol{\sigma}_2 = -i\boldsymbol{\sigma}_1 \times \boldsymbol{\sigma}_2$ and so that C_i and $\tilde{C}_i$ appear in the combination $C_i - \tilde{C}_i$ —*i.e.*, in the ZRA,

iv) $^3S_1 \to {}^3P_1$ pn: $C_1 - 3C_3 - (C_i \to \tilde{C}_i)$.

Finally, in the case of the $^3S_1$ parameter $c_t(k)$, we exploit the isotopic spin analog of eq. (85):

$$
\boldsymbol{\tau}_1 - \boldsymbol{\tau}_2 = -\frac{i}{2}(\boldsymbol{\tau}_1 \times \boldsymbol{\tau}_2)(1 + \boldsymbol{\tau}_1 \cdot \boldsymbol{\tau}_2), \qquad (86)
$$

so that, in the ZRA, the dependence must be

v) $^3S_1 \to {}^1P_1$ pn: $\tilde{C}_6 + \frac{1}{2}(C_2 - C_4)$.

---

[8] This is clear since a gradient operator acting on an S-state wave function yields a term linear in r, which vanishes at the origin.

An alternate way to understand the feature that there can be only five independent low-energy observables has recently been presented by Girlanda [52]. Using the feature that in the nonrelativisitic limit the twelve forms given in the definition of the EFT potential eq. (57) can be replaced by the twelve relativistic operators

$$
\begin{aligned}
\mathcal{O}_1 &= \bar{\psi}\gamma^\mu\psi\bar{\psi}\gamma_\mu\gamma_5\psi, \\
\tilde{\mathcal{O}}_1 &= \bar{\psi}\gamma^\mu\gamma_5\psi\partial^\nu(\bar{\psi}\sigma_{\mu\nu}\psi), \\
\mathcal{O}_2 &= \bar{\psi}\gamma^\mu\psi\bar{\psi}\tau_3\gamma_\mu\gamma_5\psi, \\
\tilde{\mathcal{O}}_2 &= \bar{\psi}\gamma^\mu\gamma_5\psi\partial^\nu(\bar{\psi}\tau_3\sigma_{\mu\nu}\psi), \\
\mathcal{O}_3 &= \bar{\psi}\tau_a\gamma^\mu\psi\bar{\psi}\tau^a\gamma_\mu\gamma_5\psi, \\
\tilde{\mathcal{O}}_3 &= \bar{\psi}\tau_a\gamma^\mu\gamma_5\psi\partial^\nu(\bar{\psi}\tau^a\sigma_{\mu\nu}\psi), \\
\mathcal{O}_4 &= \bar{\psi}\tau_3\gamma^\mu\psi\bar{\psi}\gamma_\mu\gamma_5\psi, \\
\tilde{\mathcal{O}}_4 &= \bar{\psi}\tau_3\gamma^\mu\gamma_5\psi\partial^\nu(\bar{\psi}\sigma_{\mu\nu}\psi), \\
\mathcal{O}_5 &= \mathcal{I}_{ab}\bar{\psi}\tau_a\gamma^\mu\psi\bar{\psi}\tau_b\gamma_\mu\gamma_5\psi, \\
\tilde{\mathcal{O}}_5 &= \mathcal{I}_{ab}\bar{\psi}\tau_a\gamma^\mu\gamma_5\psi\partial^\nu(\bar{\psi}\tau_b\sigma_{\mu\nu}\psi), \\
\mathcal{O}_6 &= i\epsilon_{ab3}\bar{\psi}\tau_a\gamma^\mu\psi\bar{\psi}\tau_b\gamma_\mu\gamma_5\psi, \\
\tilde{\mathcal{O}}_6 &= i\epsilon_{ab3}\bar{\psi}\tau_a\gamma^\mu\gamma_5\psi\partial^\nu(\bar{\psi}\tau_b\sigma_{\mu\nu}\psi), \quad (87)
\end{aligned}
$$

then, with the use of Fierz transformations and the free particle equation of motion, one finds the six conditions

$$
\begin{aligned}
\mathcal{O}_3 &= \mathcal{O}_1, \\
\mathcal{O}_2 - \mathcal{O}_4 &= 2\mathcal{O}_6, \\
\tilde{\mathcal{O}}_3 + 3\tilde{\mathcal{O}}_1 &= 2m_N(\mathcal{O}_1 + \mathcal{O}_3), \\
\tilde{\mathcal{O}}_2 + \tilde{\mathcal{O}}_4 &= m_N(\mathcal{O}_2 + \mathcal{O}_4), \\
\tilde{\mathcal{O}}_2 - \tilde{\mathcal{O}}_4 &= -2m_N\mathcal{O}_6 - \tilde{\mathcal{O}}_6, \\
\tilde{\mathcal{O}}_5 &= \mathcal{O}_5. \quad (88)
\end{aligned}
$$

Finally, using the feature that the operators $\mathcal{O}_6$ and $\tilde{\mathcal{O}}_6$ have the same form in the lowest-order nonrelativisitic expansion, we find an effective (pionless) potential

$$
\begin{aligned}
V_{EFT} = \frac{2\mu^2}{\Lambda_\chi^3}\Bigg[& \left(C_1 + (C_2 + C_4)\left(\frac{\tau_1 + \tau_2}{2}\right)_z + C_5\mathcal{I}_{ab}\tau_{1a}\tau_{2b}\right) \\
& \times (\boldsymbol{\sigma}_1 - \boldsymbol{\sigma}_2)\cdot\{\boldsymbol{p}_1 - \boldsymbol{p}_2, f_\mu(r)\} \\
& + i\tilde{C}_1(\boldsymbol{\sigma}_1\times\boldsymbol{\sigma}_2)\cdot[\boldsymbol{p}_1 - \boldsymbol{p}_2, f_\mu(r)] \\
& + iC_6\epsilon_{ab3}\tau_{1a}\tau_{2b}[\boldsymbol{p}_1 - \boldsymbol{p}_2, f_\mu(r)]\Bigg] \quad (89)
\end{aligned}
$$

expressed in terms of just five independent constants, as required. Of course, the use of the free particle equation of motion means that binding effects are omitted in this reduction. Binding effects arise in terms higher order in the chiral expansion, so this omission is consistent with the use of the LO form of the effective potential, as done in our analysis. The Girlanda and Zhu et al. forms are then completely equivalent and one can choose to use either form as the low-energy parity-violating potential. In this article, we shall continue to use the conventional analysis of Zhu et al., since it can be connected straightforwardly to the DDH picture, within which nearly all experimental results have been presented.

The effect of higher-order terms, which are omitted in our lowest-order expansion, can be gauged by the use of finite-range and realistic nucleon wave functions, whereby the simple dependences of Danilov parameters on Zhu coefficients expounded above are modified. For example, Desplanques and Benayoun quote the approximate results [53]

$$
\begin{aligned}
\lambda_s^{pp} = -K_p\Big[& B_6(C_1 + C_2 + C_3 + C_4 - 2C_5 + (C_i \to \tilde{C}_i)) \\
& + B_7(C_1 + C_2 + C_3 + C_4 - 2C_5 - (C_i \to \tilde{C}_i))\Big], \\
\lambda_s^{nn} = -K_p\Big[& B_6(C_1 - C_2 + C_3 - C_4 - 2C_5 + (C_i \to \tilde{C}_i)) \\
& + B_7(C_1 - C_2 + C_3 - C_4 - 2C_5 - (C_i \to \tilde{C}_i))\Big], \\
\lambda_s^{np} = -K_p\Big[& B_6(C_1 + C_3 + 4C_5 + (C_i \to \tilde{C}_i)) \\
& + B_7(C_1 + C_3 + 4C_5 - (C_i \to \tilde{C}_i))\Big], \\
\lambda_t = -K_p\Big[& B_4(C_1 - 3C_3 + (C_i \to \tilde{C}_i)) \\
& + B_5(C_1 - 3C_3 - (C_i \to \tilde{C}_i))\Big], \\
\rho_t = -K_p\Big[& B_2\left(\frac{1}{2}(C_2 - C_4) + \tilde{C}_6\right) \\
& + B_3\left(\frac{1}{2}(C_2 - C_4) - \tilde{C}_6\right)\Big], \quad (90)
\end{aligned}
$$

where $K_p = 2\Lambda_\chi^9/m_N^4 m_\rho^6$ and the Reid soft core potential values for the B_i are found to be

$$
B_i = [-0.0043, 0.0005, -0.0009, -0.0022, -0.0067, 0.0003]
$$
for $i = 2, 3, \ldots, 7$.

We see then that the size of the finite-range corrections to the lowest-order results are given by

$$
|B_3/B_2| = 0.12, \qquad |B_7/B_6| = 0.04
$$
and
$$
|B_4/B_5| = 0.41.
$$

The first two ratios are rather small and suggest that zero range is a reasonable first approximation. In the case of the ratio $|B_4/B_5|$ there are sizable corrections to the zero-range result as a consequence of important tensor force contributions.

Alternatively, for example, in a "hybrid" pionless theory which uses the pionless potential but with AV18 wave functions, choosing $m \sim m_\pi$ Liu has evaluated these cobinations and finds [49]

$$
B_i = [0.0014, 0.0008, 0.0005, 0.0008, 0.0023, 0.0003]
$$
for $i = 2, 3, \ldots, 7$,

which are somewhat different from the estimates of Desplanques and Beneyoun and indicate some of the uncertainties associated with such analyses.

In any case we see that NLO and higher-order corrections can be of order 25% or so, which is suggestive of omitted terms in the LO chiral expansion of order nuclear binding energy or Fermi momentum —$\sim 250\,\mathrm{MeV}$— over the usual chiral expansion parameter $\Lambda_\chi \sim 4\pi F_\pi \sim 1\,\mathrm{GeV}$. An important goal of future analyses should be to include such effects by proceeding beyond the simple LO analysis given herein.

We now address the form in which to present predictions of the theory. As emphasized above, in the past most experimental numbers are interpreted in terms of the DDH parameters f_π^1, $h_{\rho,\omega}^i$. However, in an effective field-theoretic framework one wants to express predictions in terms of the parameters of the theory —in our case $C_i\,\tilde{C}_i$. However, because these ten constants must appear only in the combinations given above in analysis of threshold processes, it is more convenient to represent all predictions in terms of the five Danilov parameters, which have a rather direct connection to observables. Before presenting these predictions, however, we first show how these five parameters can be (approximately) analytically connected to the underlying low-energy constants. As an example, consider the parameter λ_t. Since the associated interaction is short-ranged, we can use this feature in order to simplify the analysis. For example, we can determine the shift in the deuteron wave function associated with parity violation by demanding orthogonality with the $^3S_1$ scattering state, which yields, using the simple asymptotic form of the bound-state wave function [54,55]

$$\psi_d(r) = [1 + \rho_t(\boldsymbol{\sigma}_p + \boldsymbol{\sigma}_n)\cdot -i\boldsymbol{\nabla} + \lambda_t(\boldsymbol{\sigma}_p - \boldsymbol{\sigma}_n)\cdot -i\boldsymbol{\nabla}]$$
$$\times \sqrt{\frac{\gamma}{2\pi}}\frac{1}{r}e^{-\gamma r}, \qquad (91)$$

where $E = -\gamma^2/M = -2.23\,\mathrm{MeV}$ is the deuteron binding energy. Now the shift generated by $V^{PV}(\boldsymbol{r})$ is found to be [54,55]

$$\delta\psi_d(\boldsymbol{r}) \simeq \int \mathrm{d}^3r' G(\boldsymbol{r},\boldsymbol{r}') V^{PV}(\boldsymbol{r}')\psi_d(\boldsymbol{r}')$$
$$= -\frac{m_N}{4\pi}\int \mathrm{d}^3r' \frac{e^{-\gamma|\boldsymbol{r}-\boldsymbol{r}'|}}{|\boldsymbol{r}-\boldsymbol{r}'|}V^{PV}(\boldsymbol{r}')\psi_d(\boldsymbol{r}')$$
$$\simeq \frac{m_N}{4\pi}\boldsymbol{\nabla}\left(\frac{e^{-\gamma r}}{r}\right)\cdot\int \mathrm{d}^3r'\,\boldsymbol{r}'V^{PV}(\boldsymbol{r}')\psi_d(\boldsymbol{r}'), \quad (92)$$

where the last step is permitted by the short range of $V^{PV}(\boldsymbol{r}')$. Comparing eqs. (92) and (91) yields then the identification

$$\sqrt{\frac{\gamma}{2\pi}}\lambda_t\chi_t \equiv i\frac{m_N}{16\pi}\xi_0^\dagger\int \mathrm{d}^3r'(\boldsymbol{\sigma}_1-\boldsymbol{\sigma}_2)\cdot\boldsymbol{r}'V^{PV}(\boldsymbol{r}')\psi_d(\boldsymbol{r}')\chi_t\xi_0, \qquad (93)$$

where we have included the normalized isoscalar wave function ξ_0 since the potential involves $\boldsymbol{\tau}_1$, $\boldsymbol{\tau}_2$. When operating on such an isosinglet np state the PV potential can be written as

$$V^{PV}(\boldsymbol{r}') = \frac{2}{\Lambda_\chi^3}[(C_1 - 3C_3)(\boldsymbol{\sigma}_1 - \boldsymbol{\sigma}_2)$$
$$\cdot(-i\boldsymbol{\nabla}f_m(r) + 2f_m(r)\cdot -i\boldsymbol{\nabla})$$
$$+(\tilde{C}_1 - 3\tilde{C}_3)(\boldsymbol{\sigma}_1 \times \boldsymbol{\sigma}_2)\cdot\boldsymbol{\nabla}f_m(r)], \qquad (94)$$

whereby eq. (93) becomes

$$\sqrt{\frac{\gamma}{2\pi}}\lambda_t\chi_t \simeq \frac{2m_N}{16\pi\Lambda_\chi^3}\frac{4\pi}{3}(\boldsymbol{\sigma}_1-\boldsymbol{\sigma}_2)^2\chi_t\int_0^\infty \mathrm{d}r\,r^3$$
$$\times \left[-2(3C_3 - C_1)f_m(r)\frac{\mathrm{d}\psi_d(r)}{\mathrm{d}r}\right.$$
$$\left.+ (3\tilde{C}_3 - 3C_3 - \tilde{C}_1 + C_1)\frac{\mathrm{d}f_m(r)}{\mathrm{d}r}\psi_d(r)\right] =$$
$$\sqrt{\frac{\gamma}{2\pi}}\cdot 4\chi_t\frac{1}{12}\frac{4m_N m^3}{4\pi\Lambda_\chi^3}\frac{K_c}{(\gamma+m)^2}, \qquad (95)$$

where

$$K_c = 2m(6C_3 - 3\tilde{C}_3 - 2C_1 + \tilde{C}_1)$$
$$+ \gamma(15C_3 - 3\tilde{C}_3 - 5C_1 + \tilde{C}_1), \qquad (96)$$

or

$$\lambda_t \simeq -\frac{m_N m^3}{3\pi\Lambda_\chi^3}\frac{K_c}{(\gamma+m)^2}. \qquad (97)$$

Performing the indicated integration and using $m \sim m_\rho$ we find the result

$$\lambda_t \simeq -0.020(-2C_1 + \tilde{C}_1) + 0.060(-2C_3 + \tilde{C}_3). \qquad (98)$$

However, this is clearly an overestimate because it was obtained i) using the asymptotic form of the wave function and ii) omits short-range correlation effects. In order to deal approximately with the short-distance properties of the deuteron wave function, we modify the exponential form to become constant inside the deuteron radius R [54,55]

$$\sqrt{\frac{\gamma}{2\pi}}\frac{1}{r}e^{-\gamma r} \to N\begin{cases}\frac{1}{R}e^{-\gamma R}, & r \le R, \\ \frac{1}{r}e^{-\gamma r}, & r > R,\end{cases} \qquad (99)$$

where

$$N = \sqrt{\frac{\gamma}{2\pi}}\frac{\exp\gamma R}{\sqrt{1 + \frac{2}{3}\gamma R}}$$

is the modified normalization factor and we use $R = 1.6\,\mathrm{fm}$. As to the short-range (Jastrow) correlation, we multiply the wave function by the simple phenomenological form [56]

$$\phi(r) = 1 - ce^{-dr^2}, \quad \text{with} \quad c = 0.6,\; d = 3\,\mathrm{fm}^{-2}. \quad (100)$$

With these modifications we determine the much more reasonable values for the Danilov parameter λ_t

$$\lambda_t = [0.003(-2C_3 + \tilde{C}_3) - 0.002(-2C_1 + \tilde{C}_1)]m_N^{-1}. \quad (101)$$

In this way approximate analytic forms can also be found for the remaining Danilov parameters [57].

However, it is obviously preferable to use estimates obtained using the best available wave functions. In this way Liu determines that [49]

$$\lambda_t = [0.0045(-2.23C_3 + \tilde{C}_3) - 0.0015(-2.23C_1 + \tilde{C}_1)]m_N^{-1}. \qquad (102)$$

The similarity with the approximate analytic expression is obvious —the discrepancy with the coefficients involving C_3, $\tilde{C}_3$ is again due to effects from the tensor interaction— and the other coefficients are found in this way to be

$$\lambda_s^{pp} = 0.0043[(C_1 + C_2 + C_3 + C_4 - 2C_5)$$
$$+ 1.27(C_i \to \tilde{C}_i)]m_N^{-1},$$
$$\lambda_s^{nn} = 0.0046[(C_1 - C_2 + C_3 - C_4 - 2C_5)$$
$$+ 1.22(C_i \to \tilde{C}_i)]m_N^{-1},$$
$$\lambda_s^{np} = 0.0047[(C_1 + C_3 + 4C_5) + 1.24(C_i \to \tilde{C}_i)]m_N^{-1}$$
$$\rho_t = 0.0031[\tilde{C}_6 + 0.60(C_2 - C_4)]m_N^{-1}. \quad (103)$$

A connection between the underlying EFT Lagrangian and the empirical Danilov parameters has thus been established.

In the next section we shall describe how the results of various low-energy experiments can be expressed in terms of the Danilov parameters. In the design of such experiments, it is obviously useful to have at hand numerical values for the size of these quantities and the use of the DDH estimates for C_i, $\tilde{C}_i$ provides a reasonable way to provide such numbers. Of course, the completely consistent and correct way to accomplish this is to use a pionful theory, with one- and two-pion exchange pieces described in terms of f_π^1 and the remaining terms written in terms of short-distance quantities C_i, $\tilde{C}_i$. However, for the reasonable and simple estimates needed below we shall instead employ approximate values for the Danilov parameters which include the effects of heavy meson exchange for λ_t, $\lambda_s^{0,1,2}$ and pion exchange for the parameter ρ_t, yielding

$$^{DDH}\lambda_s^{pp} = 2.3 \times 10^{-7} m_N^{-1},$$
$$^{DDH}\lambda_s^{nn} = 2.1 \times 10^{-7} m_N^{-1},$$
$$^{DDH}\lambda_s^{np} = 0.8 \times 10^{-7} m_N^{-1},$$
$$^{DDH}\lambda_t = 0.6 \times 10^{-7} m_N^{-1},$$
$$^{DDH}\rho_t = 5.6 \times 10^{-7} m_N^{-1}. \quad (104)$$

These numbers, of course, should *not* be treated as being in any sense precise. However, they are useful in estimating the possible size of experimental effects, as will be seen below.

4 Experimental program

Having developed a connection of the five Danilov parameters with the underlying effective Lagrangian, our next task is to develop a program whereby experimental values of these quantities can be reliably determined. Since we desire to generate *definitive* values for these constants, we certainly do not wish to introduce nuclear-physics uncertainties into the analysis. Thus we shall require *only* experiments involving systems with $A \leq 4$, for which nuclear wave functions are well determined. We recognize that, by imposing this requirement we eliminate the opportunity to enhance experimental signals via the careful

choice of near-degenerate opposite parity levels that has permitted experiments in ^{18}F [45], ^{19}F [44], ^{21}Ne [58] with precision of parts in 100000 or even in 10000 to provide useful input into the parity-violating interaction puzzle. As a consequence, the experimental signals we need to analyze will be a part in 10000000 or even smaller! Nevertheless, we consider this a price worth paying in order to have confidence in the interpretation of the experimental signals.

Since there is a need to determine *five* parameters, we clearly require a minimum of *five* independent measurements. To the extent that all experiments are performed at threshold, the analysis can only involve five independent combinations of EFT parameters. We elect to present our predictions in terms of the five Danilov coefficients, because of their phenomenological signifcance, but this choice is somewhat arbitrary and there is no implication that this representation is superior to other possibilities. (Of course, if we stray above threshold, additional terms can come in.) As discussed above, we do not possess at this time a modern first principles theoretical analysis of each of the experimental possibilities. Hence, in the discussion below we present approximate existing estimates within the simple pionless theory, with an effective contact potential in terms of the Danilov coefficients. A fully consistent pionless calculation would then evaluate corresponding diagrams generated within this framework. However, this does not exist at this time, so that in order to match onto experiment we shall have to use results from a variety of different calculational schemes. *Providing a rigorous theoretical analysis of each experiment within the same rigorous calculational framework should be a priority for future work.* With these caveats, we shall consider five such possible reactions in turn:

i) **pp scattering asymmetry**: the first and simplest such reaction has already been performed and involves the asymmetry in the scattering cross-section for longitudinally polarized protons on an unpolarized proton target. The experimental signal is the difference in the right- and left-handed total scattering cross-sections divided by their sum

$$A_h = \frac{\sigma_+ - \sigma_-}{\sigma_+ + \sigma_-}.$$

Results are available from a Bonn experiment at lab energy 13.6 GeV and from a PSI experiment at 45 MeV[9]:

$$A_h(13.6\,\text{MeV}) = -(0.93 \pm 0.20 \pm 0.05) \times 10^{-7} \, [60],$$
$$A_h(45\,\text{MeV}) = -(1.57 \pm 0.23) \times 10^{-7} \, [61]. \quad (106)$$

---

[9] Note that there also exists a Los Alamos measurement at 15 MeV

$$A_h(15\,\text{MeV}) = -(1.7 \pm 0.8) \times 10^{-7} \, [59] \quad (105)$$

which is quite consistent with the asymmetry measured at 13.6 MeV. However, because of its superior precision, we shall use only the Bonn result.

The feature that the PSI number is about 60% greater than its Bonn analog is consistent with the feature that the asymmetry should depend roughly linearly on the proton momentum which depends on energy as

$$\frac{k_{PSI}}{k_{Bonn}} \sim \sqrt{\frac{45}{13.6}} = 1.8. \quad (107)$$

Thus these two experiments should *not* be considered as yielding independent numbers.

The connection with the Danilov parameters can be found by calculating the helicity-correlated cross-sections which, since the initial state must be in a spin-singlet, must have the form [57]

$$\sigma_\pm = \int d\Omega \frac{1}{2} \operatorname{Tr} \mathcal{M}(\boldsymbol{k}', \boldsymbol{k}) \frac{1}{2}(1 + \boldsymbol{\sigma}_2 \cdot \hat{k})$$
$$\cdot \frac{1}{4}(1 - \boldsymbol{\sigma}_1 \cdot \boldsymbol{\sigma}_2)\mathcal{M}^\dagger(\boldsymbol{k}', \boldsymbol{k}) =$$
$$|m_s(k)|^2 \pm 4k \operatorname{Re} m_s^*(k)d_s^{nn}(k) + \mathcal{O}(d_s^2). \quad (108)$$

Defining the asymmetry via the sum and difference of such helicity cross sections and neglecting the tiny P-wave scattering, we have then

$$A_h(E_{\text{threshold}}) = \frac{\sigma_+ - \sigma_-}{\sigma_+ + \sigma_-}$$
$$= -\frac{8k \operatorname{Re} m_s^*(k)d_s^{nn}(k)}{2|m_s(k)|^2} = -4k\lambda_s^{pp}. \quad (109)$$

Thus the threshold helicity-correlated pp scattering asymmetry provides a direct measure of the parity-violating parameter λ_s^{pp}.

Of course, the actual experiments are performed not at threshold but rather at the finite laboratory energies quoted above. Converting to momentum we find that the corresponding numbers are $160 \, \mathrm{MeV}/c$ and $290 \, \mathrm{MeV}/c$ respectively, so that the simple threshold relation above must be modified and the relation of the asymmetry to the Danilov parameter λ_s^{pp} becomes somewhat more complex [49]:

$$A_h(13.6 \, \mathrm{MeV}) = -0.449 m_N \lambda_s^{pp},$$
$$A_h(45 \, \mathrm{MeV}) = -0.795 m_N \lambda_s^{pp}. \quad (110)$$

We see that the dominant dependence is on the Danilov parameter λ_s^{pp}, with corrections relatively small at $45 \, \mathrm{MeV}$ and tiny at $13.6 \, \mathrm{MeV}$. However, the momentum involved in the $45 \, \mathrm{MeV}$ experiment certainly gives one pause and a recent careful EFT analysis by Phillips *et al.* suggests that the agreement between the simple momentum-scaled experimental numbers may be fortuitous, as NLO corrections are expected to be significant [62].

It should be noted for completeness that there exist additional measurements of the pp asymmetry at $221.3 \, \mathrm{MeV}$ and at $800 \, \mathrm{MeV}$ performed at TRIUMF and LANL, respectively. In the case of the former, the energy was carefully selected in order that S-P parity mixing effects vanish, leaving sensitivity to P-D mixing and allowing separation of the DDH parameters involving isoscalar rho exchange and omega exchange. A precise number [63]

$$A_h(221.3 \, \mathrm{MeV}) = (0.83 \pm 0.29 \pm 0.17) \times 10^{-7} \quad (111)$$

was obtained. However, P-D mixing is beyond the scope of our parity-violating EFT and would have to be accounted for by inclusion of a an entirely new set of phenomenological parameters. Nevertheless Carlson *et al.* have reported being able to fit both the low- and higher-energy data within the DDH scheme [64]. The $800 \, \mathrm{MeV}$ result [65]

$$A_h(800 \, \mathrm{MeV}) = (2.4 \pm 1.1) \times 10^{-7} \quad (112)$$

is positive, as expected from the feature that the S-P and P-D interference terms both contribute positively above $220 \, \mathrm{MeV}$, but again a detailed analysis requires input which is well beyond the scope of our low-energy EFT methods.

Considering only the low-energy results then, we find from the above measurements the result

$$\lambda_s^{pp} \simeq (2.0 \pm 0.3) \times 10^{-7} m_N^{-1} \quad (113)$$

which is the only really solid experimental measurement of a Danilov parameter which exists at present. Note that this limit is quite consistent with the DDH "best value" estimate

$$^{DDH}\lambda_s^{pp} = 2.3 \times 10^{-7} m_N^{-1}. \quad (114)$$

ii) $p\alpha$ scattering asymmetry: a second experiment which is relevant to this program is a $45 \, \mathrm{MeV}$ proton helicity asymmetry experiment on a ^{4}He target, which was performed at PSI, yielding [43]

$$A_h(45 \, \mathrm{MeV}) = -(3.3 \pm 0.9) \times 10^{-7}. \quad (115)$$

The problem here is that we do not yet have a precise theoretical prediction in terms of Danilov parameters. There does exist, however, a Desplanques and Missimer calculation [50]

$$A_h(45 \, \mathrm{MeV}) = -\left[0.48\left(\lambda_s^{pp} + \frac{1}{2}\lambda_s^{pn}\right) + 1.07\left(\rho_t + \frac{1}{2}\lambda_t\right)\right]m_N \quad (116)$$

which provides a constraint

$$0.48\lambda_s^{pn} + 2.14\left(\rho_t + \frac{1}{2}\lambda_t\right) = (4.6 \pm 2.0) \times 10^{-7} m_N^{-1} \quad (117)$$

quite consistent with the DDH "best value" estimate

$$0.48^{DDH}\lambda_s^{pn} + 2.14\left(^{DDH}\rho_t + \frac{1}{2}{}^{DDH}\lambda_t\right) = 4.8 \times 10^{-7} m_N^{-1}. \quad (118)$$

However, what is needed here is a *definitive* theoretical analysis. (Of course, the rather deep 28 MeV binding energy of ^{4}He might give one pause as to whether a calculation in terms of simply the threshold (Danilov) parameters is adequate. On the other hand, a recent paper is successful in explaining a correlation between the triton and alpha binding energies in various pionless theories [66], so this issue invites further study.)

An additional source of information is provided by experiments involving the radiative capture of neutrons on a proton target

$$n + p \to d + \gamma$$

for which a solid theoretical analysis *is* available. There exist two independent parity-violating observables in this reaction:

iii) Asymmetry in *np* capture: one possible measurement involves the photon asymmetry in the case of polarized neutron capture, for which one finds [49]

$$A_\gamma = -0.093 m_N \rho_t. \tag{119}$$

On the experimental side, there exist already two results, one from an old Grenoble measurement and one from a new LANSCE experiment:

$$\begin{aligned} A_\gamma &= (0.6 \pm 2.1) \times 10^{-7} \ [67], \\ A_\gamma &= (-1.1 \pm 2.0 \pm 0.2) \times 10^{-7} \ [68]. \end{aligned} \tag{120}$$

While these two numbers are in good agreement and are of impressive precision, considerable improvement is still called for. That is because the dominant piece of the Danilov parameter ρ_t comes from one-pion exchange and therefore depends upon the PV pion emission amplitude f_π^1. Using the DDH "best value" for this number, the corresponding experimental prediction for A_γ is

$$^{DDH}A_\gamma^{th} = -5 \times 10^{-8} \tag{121}$$

which is a full order of magnitude *smaller* than the levels probed by the existing experiments! For this reason, the LANSCE experiment has been disassembled and moved to the new fundamental neutron physics beamline at SNS, where the associated increased intensity should allow a measurement at the level of a few parts per billion. In fact this experiment is the commissioning experiment for this beamline and should commence later this year.

The theoretical prediction is in good agreement with previous calculations —cf. [69]— and depends predominantly on the parity-violating pion coupling f_π^1. Thus measurement of the $np \to d\gamma$ asymmetry with the hoped for precision should finally resolve the burning question of whether this long-range coupling is of the order or considerably smaller than its DDH prediction.

iv) Circular polarization in *np* capture: an independent probe of parity violation in radiative neutron capture

is provided by the possibility of measuring the circular polarization of the outgoing photon resulting from the capture of *unpolarized* neutrons, for which the prediction in terms of Danilov parameters is [49]

$$P_\gamma = -0.161 m_N \lambda_s^{np} + 0.670 m_N \lambda_t. \tag{122}$$

This is an old idea and the first attempt to measure this parameter was done in 1972 by Lobashov *et al.* who reported a value

$$P_\gamma = -(1.3 \pm 0.45) \times 10^{-6} \ [70]. \tag{123}$$

It was later realized that this experiment was contaminated by polarized bremsstrahlung photons from fission products in the reactor and the number was subsequently revised downward to

$$P_\gamma = (1.8 \pm 1.8) \times 10^{-7} \ [71]. \tag{124}$$

Again, however, despite its impressive precision, a considerably improved measurement is needed, since use of eq. (122) with DDH "best values" for the Danilov parameters yields a prediction

$$^{DDH}P_\gamma^{th} = 2.7 \times 10^{-8} \tag{125}$$

considerably *below* the current experimental precision. Improvement of the existing limit will be challenging, however, because of the relatively low efficiencies of circular polarization detectors, and it may be advantageous to use the time-reversed reaction

$$\gamma + d \to n + p$$

for which the asymmetry using circularly polarized photons is equal, using detailed balance, to the circular polarization in the radiative capture reaction. Nevertheless, either experiment will be extraordinarily difficult since the theoretical expectation is so small.

Note that the predicted value depends only on the short-distance–dominated Danilov parameters λ_s^{np} and λ_t and is independent of the PV pion coupling. Nevertheless the predicted DDH value is in reasonable agreement with previous estimates —cf. [69].

v) As a fifth experiment in this program, one can utilize neutron spin rotation when passing through a parahydrogen target, for which the rotation rate is predicted to be [49]

$$\frac{d\phi^{np}}{dz} = [2.500\lambda_s^{np} - 0.571\lambda_t + 1.412\rho_t]m_N \text{ rad/m.} \tag{126}$$

The use of the DDH "best value" numbers then predicts the small number

$$^{DDH}\left(\frac{d\phi^{np}}{dz}\right)^{th} = 9.6 \times 10^{-7} \text{ rad/m.} \tag{127}$$

but a planned experiment at SNS anticipates a precision at the level of 2.7×10^{-7} rad/m and will provide an important data point. However, such experiments are very challenging, since one must shield the system from external magnetic fields, for which Faraday rotation in the Earth's magnetic field yields a rotation considerably larger than those being sought due to the weak interactions.

The theoretical prediction here is of the same rough size and sign as that given in [72] and [49] but differs in sign from an earlier prediction —[73].

We see then that in principle there do indeed exist a complete set of independent low-energy measurements which could be utilized in order to determine the five Danilov parameters. However, since each of the experiments is so challenging it is certainly advisable to *overdetermine* these quantities by performing additional parity-violating experiments in $A < 4$ systems. There are a number of possibilities here.

a) Neutron spin rotation on ^{4}He: this is an experiment which is already underway at NIST. As in the case of $p\alpha$ scattering the use of ^{4}He and its ~ 28 MeV binding energy means that the use of EFT methods may be a bit of a stretch. Also, a definitive calculation of the rotation angle has not been performed. Nevertheless an estimate

$$\frac{\mathrm{d}\phi^{n\alpha}}{\mathrm{d}z} = [0.60\lambda_s^{np} + 1.34\lambda_t - 2.68\rho_t + 1.2\lambda_s^{nn}]\, m_N\,\mathrm{rad/m} \tag{128}$$

is available [50]. The use of the DDH estimates for the Danilov parameters yields then

$$^{DDH}\left(\frac{\mathrm{d}\phi^{n\alpha}}{\mathrm{d}z}\right)^{th} = -11.7 \times 10^{-7}\,\mathrm{rad/m},$$

which is larger than and of opposite sign compared to the corresponding np number quoted above. There exists an experimental number for this quantity from a University of Washington Thesis [74]

$$\left(\frac{\mathrm{d}\phi^{n\alpha}}{\mathrm{d}z}\right)^{exp} = (8 \pm 14) \times 10^{-7}. \tag{129}$$

However, it is clear that the precision of this measurement is not high enough to place significant limits on the Danilov parameters.

b) Radiative nd capture —$n + d \to t + \gamma$— is being considered at SNS as a possible followup experiment to the radiative np capture. Again a definitive calculation of the photon asymmetry has not yet been performed. However, an estimate

$$A_\gamma^n = [1.35\rho_t + 0.58\lambda_s^{nn} + 1.15\lambda_t + 0.50\lambda_s^{pn}]m_N \tag{130}$$

has been given [53]. Using the DDH "best value" estimates, this yields an effect

$$^{DDH}A_\gamma^{nth} = 9.9 \times 10^{-7} \tag{131}$$

much larger than the corresponding np value. However, the existing experimental number [75]

$$A_\gamma^{n\,exp} = (4.2 \pm 3.8) \times 10^{-6} \tag{132}$$

will have to be improved by nearly an order of magnitude in order to say something meaningful.

Another possibility is to measure the photon asymmetry following the capture of an unpolarized neutron by a polarized deuteron, for which one finds [53]

$$A_\gamma^d = -[3.56\rho_t + 0.24\lambda_s^{nn} + 1.39\lambda_t + 0.71\lambda_s^{pn}]m_N \tag{133}$$

yielding an even larger signal using the DDH "best value" estimates

$$^{DDH}A_\gamma^{d\,th} = 2.2 \times 10^{-6}. \tag{134}$$

However, a high-polarization deuterium target would be required.

Finally, one can imagine measuring the circular polarization of the photon following the capture of an unpolarized neutron, for which an estimate

$$P_\gamma = -[2.73\rho_t + 0.57\lambda_s^{nn} + 1.56\lambda_t + 0.73\lambda_s^{pn}]m_N \tag{135}$$

has been given [53]. Using the DDH "best value" numbers we find an estimate

$$^{DDH}P_\gamma^{th} = -1.8 \times 10^{-6} \tag{136}$$

again much larger than its corresponding $np \to d\gamma$ value. However, the efficient detection of circular polarization represents a challenge, and the reverse reaction $\gamma + t \to n + d$ is associated with significant safety issues because of the need for a tritium target and is probably not a serious consideration.

c) pd scattering, for which at 15 MeV has the longitudinal asymmetry [50]

$$A_L^{pd} = -[0.21\rho_t + 0.07\lambda_s^{pp} - 0.13\lambda_t - 0.04\lambda_s^{pn}]m_N. \tag{137}$$

This calculated value is based on the Desplanques-Missimer/Bethe-Goldstone estimate, and should be updated with a modern three-body calculation. However, the use of the DDH "best values" indicates an effect of the size

$$^{DDH}A_L^{th} = -1.3 \times 10^{-7}. \tag{138}$$

This experiment has been performed both at LANL at 15 MeV [76] and at PSI at 45 MeV [77]. However, the measured asymmetry is available only over a limited range of angles. Also the experiments do not distinguish elastic and breakup events. Thus, a detailed theoretical analysis would be required in order to extract information from the existing numbers.

d) Another possibility being considered is neutron spin rotation on deuterium, although experimentally this presents a number of challenges. However, a new precision theoretical estimate is available [78]:

$$\frac{1}{\rho}\frac{\mathrm{d}\phi}{\mathrm{d}z} = 2\frac{m_\pi^3}{\Lambda_\chi^3}[270\tilde{C}_6 + 3.6C_1 - 0.1\tilde{C}_1 - 0.5(C_2 + C_4)] \tag{139}$$

in terms of the effective potential developed by Girlanda —eq. (89). Using a liquid-deuterium density of 0.4×10^{23} atoms/cm^3 one finds a "best value" predicted size of about 5×10^{-6} rad/m which is about an order of magnitude larger than the corresponding np number and thus should be seriously considered as a possible source of information provided the experimental challenges can be overcome.

e) An additional followup experiment at SNS is a measurement of the proton asymmetry in the capture of polarized neutrons by ^{3}He —$\boldsymbol{n}^3$He $\rightarrow pt$. An estimate by M. Viviani has been provided within the DDH model [79]:

$$A_p = -0.18 f_\pi^1 - 0.14 h_\rho^0 + 0.27 h_\rho^1 + 0.0012 h_\rho^2$$
$$-0.13 h_\omega^0 + 0.05 h_\omega^1$$

and use of best values yields the estimate $A_p^{\text{best value}} = 1 \times 10^{-7}$, which involves a considerable cancellation between f_π^1 and h_ρ^0 couplings. R&D is now taking place at LANSCE for such an experiment, to begin in 2011.

The completion of the five core experiments supplemented by one or more of the additional possibilities outlined above would (at last!) provide a solid base of empirically determined PV parameters.

5 Future initiatives

At the present time, we have results for only two of the five necessary experimental results. Therefore it is unknown whether implementation of EFT methods will be able to resolve the inconsistencies which exist in the current DDH analysis of hadronic PV experiments. However, through successful completion and analysis of a set of experiments such as those described above, we can anticipate obtaining a consistent set of Danilov parameters at some point is the (near?) future. An obvious question is: what happens next? To some extent the answer to this question depends on whether the results of the experimental program are in some sense surprising in that they are strongly discrepant with the DDH analysis. Let us suppose that this is *not* the case. Then a number of obvious steps are suggested:

i) Firstly, it will be interesting to determine if the values of the parameters $\tilde{C}_i / C_i$ differ from their vector dominance values suggested via single-meson exchange. This will not be easy to do, however, in that in order to make this determination one will have to go above the threshold region in order to separate matrix elements involving the commutator —$[-i\boldsymbol{\nabla}, f_m(r)]$— from those involving the anticommutator —$\{-i\boldsymbol{\nabla}, f_m(r)\}$. This analysis must be done carefully so that P-D mixing effects are appropriately included.

ii) Another interesting topic is the size of the PV coupling f_π^1, for which the present DDH-based analysis indicates a value considerably smaller than the DDH best estimate from analysis of experiments involving ^{18}F

but a value considerably larger than the DDH best estimate from analysis of experiments involving ^{133}Cs. It is always possible that the value determined from the Danilov analysis will agree with neither, but if the new number is consistent with either of the present values, something important will be learned about nuclear effects from the analysis of the "losing" experiment.

iii) Once a fully consistent set of values is obtained for the low-energy constants C_i, $\tilde{C}_i$ it will be important to see if the numbers obtained experimentally can be predicted from purely theoretical considerations. At the simplest level one can compare with the DDH expectations. However, because of the uncertainties inherent in the DDH numbers, this may be a challenge. More fundamental should be an attempt to calculate such couplings via lattice methods. Because these are two-nucleon matrix elements involving both the strong *and* weak interactions this will not be a simple calculation but is a necessary ingredient to any real understanding of hadronic parity violation. Some of the challenges associated with any such calculation have recently been discussed by Beane and Savage in the context of a lattice calculation of the pion-nucleon coupling constant [80].

6 Conclusions

The field of hadronic parity violation began in 1957 with the experiment by Tanner looking for the PV ^{19}F$(p,\alpha)^{16}$O reaction. More than fifty years (and many experiments) later we still do not have a comprehensive understanding of the PV NN interaction. Since 1980 nearly all such experiments have been analyzed within the DDH (single-meson exchange) picture, but it has been difficult to resolve the issue of a small PV pionic coupling f_π^1 indicated by measurements of the circular polarization of the photon emitted in the decay of the 1.089 MeV level of ^{18}F with that indicated by the ^{133}Cs anapole moment measurement, as well as others. In order to resolve these issues and to remove nuclear-physics uncertainties from the analysis, an effective field theory approach to the subject together with an experimental program utilizing only $A \leq 4$ systems have been developed. Both the theoretical and experimental programs were described above.

However, significant challenges remain. On the theoretical side it is important to develop state-of-the-art calculations which relate empirical results to the underlying theoretical basis. This is especially important for those experiments involving ^{4}He targets. Experimentally, the price that is paid for use of light nuclear systems is the loss of the possibility of nuclear enhancement, meaning that experiments must be done to a precision of a part in 10^8 or better. Nevertheless, this can and *must* be achieved in order to bring understanding to this field. The use of effective field-theoretic methods means that at some point in the near future we will be able to converge on a consistent set of empirical parameters —the Danilov coefficients—

which can characterize the low-energy PV NN interaction. Once this is accomplished the focus can shift to the use of these numbers to understand the previous nuclear experiments and to the theoretical prediction of such numbers from fundamental theory —QCD. Only then can we say that, after more than a half century of effort, the problem of hadronic parity violation is finally solved.

This work was supported in part by the National Science Foundation under award PHY05-53304. Many useful conversations with C.-P. Liu, M.J. Ramsey-Musolf, and S.-L. Zhu are gratefully acknowledged.

References

1. See, *e.g.*, S.C. Pieper, R.B. Wiringa, Annu. Rev. Nucl. Part. Sci. **51**, 53 (2001); B.R. Barrett, P. Navrátil, W.E. Ormand, J.P. Vary, Acta Phys. Pol. B **33**, 297 (2002).

2. See, *e.g.*, U. van Kolck, P.F. Bedaque, Annu. Rev. Nucl. Part. Sci. **52**, 339 (2002); U. van Kolck, Nucl. Phys. A **699**, 33 (2002); E. Epelbaum, Prog. Part. Nucl. Phys. **57**, 654 (2006); E. Epelbaum, H.-W. Hammer, U.-G. Meissner, arXiv:0811.1338, submitted to Rev. Mod. Phys.

3. S. Weinberg, Phys. Lett. B **251**, 288 (1990); Nucl. Phys. B **363**, 3 (1991).

4. C. Ordóñez, U. van Kolck, Phys. Lett. B **291**, 459 (1992); C. Ordóñez, L. Ray, U. van Kolck, Phys. Rev. C **53**, 2086 (1996); N. Kaiser, R. Brockmann, W. Weise, Nucl. Phys. A **625**, 758 (1997); N. Kaiser, S. Gerstendorfer, W. Weise, Nucl. Phys. A **637**, 395 (1998); N. Kaiser, Phys. Rev. C **65**, 017001 (2002) and references therein.

5. U. van Kolck, Phys. Rev. C **49**, 2932 (1994); J.L. Friar, D. Hüber, U. van Kolck, Phys. Rev. C **59**, 53 (1999).

6. E. Epelbaum, W. Glöckle, U.-G. Meißner, Nucl. Phys. A **671**, 295 (2000); Eur. Phys. J. A **19**, 125; 401 (2004); D.R. Entem, R. Machleidt, Phys. Rev. C **66**, 014002 (2002); **68**, 041001 (2003).

7. E. Epelbaum, A. Nogga, W. Glöckle, H. Kamada, U.-G. Meißner, H. Witała, Eur. Phys. J. A **15**, 543 (2002); Phys. Rev. C **66**, 064001 (2002).

8. U. van Kolck, Nucl. Phys. A **645**, 273 (1999); **787**, 405 (2007); S. Beane, P. Bedaque, M.J. Savage, U. van Kolck, Nucl. Phys. A **700**, 377 (2002); M. Birse, Phys. Rev. C **77**, 047001 (2008).

9. J.-W. Chen, G. Rupak, M.J. Savage, Nucl. Phys. A **653**, 386 (1999); S. Beane, M. Savage, Nucl. Phys. A **694**, 511 (2001); **717**, 104 (2003).

10. P.F. Bedaque, U. van Kolck, Phys. Lett. B **428**, 221 (1998); P.F. Bedaque, H.-W. Hammer, U. van Kolck, Phys. Rev. C **58**, 641 (1998); F. Gabbiani, P.F. Bedaque, H.W. Grießhammer, Nucl. Phys. A **675**, 601 (2000).

11. P.F. Bedaque, H.-W. Hammer, U. van Kolck, Nucl. Phys. A **676**, 357 (2000); P.F. Bedaque, G. Rupak, H.W. Grießhammer, H.-W. Hammer, Nucl. Phys. A **714**, 589 (2003).

12. J.-W. Chen, G. Rupak, M.J. Savage, Phys. Lett. B **464**, 1 (1999); G. Rupak, Nucl. Phys. A **678**, 405 (2000).

13. F. Ravndal, X. Kong, Phys. Rev. C **64**, 044002 (2001); M. Butler, J.-W. Chen, Phys. Lett. B **520**, 87 (2001).

14. N. Tanner, Phys. Rev. **107**, 1203 (1957).

15. R.J. Blin-Stoyle, Phys. Rev. **118**, 1605 (1960); **120**, 181 (1960).

16. F.C. Michel, Phys. Rev. B **133**, 329 (1964).

17. B. Desplanques, J.F. Donoghue, B.R. Holstein, Ann. Phys. (N.Y.) **124**, 449 (1980).

18. S.-L. Zhu, C.M. Maekawa, B.R. Holstein, M.J. Ramsey-Musolf, U. van Kolck, Nucl. Phys. A **748**, 435 (2005).

19. G.S. Danilov, Phys. Lett. **18**, 40 (1965); Phys. Lett. B **35**, 579 (1971); Sov. J. Nucl. Phys. **14**, 443 (1972).

20. E. Merzbacher, *Quantum Mechanics* (Wiley, New York, 1999).

21. Note that we are using here the standard convention in nuclear physics wherein the scattering length associated with a weak attractive potential is negative. It is common in particle physics to define the scattering length as the negative of that given here so that a weak attractive potential is associated with a positive scattering length.

22. G.S. Danilov, in *Proceedings of the XI LNPI Winter School, Leningrad, 1976*, Vol. **1**, p. 203.

23. U. van Kolck, Nucl. Phys. A **645**, 273 (1999).

24. J.D. Jackson, J.M. Blatt, *The Interpretation of Low Energy Proton-Proton Scattering*, Rev. Mod. Phys. **22**, 77 (1950).

25. See, *e.g.*, M.A. Preston, *Physics of the Nucleus* (Addison-Wesley, Reading, MA, 1962) appendix B.

26. See, *e.g.*, D.B. Kaplan, M.J. Savage, M.B. Wise, Phys. Lett. B **424**, 390 (1998); Nucl. Phys. B **534**, 329 (1998).

27. D. Phillips, Czech J. Phys. **52**, B49 (2002).

28. G.P. Lepage, *How to Renormalize the Schrödinger Equation, Nuclear Physics: Proceedings of the VIII Jorge André Swieca Summer School, Campós do Jordaó, Brazil 1997* (World Scientific, 1997) p. 135, arXiv:nucl-th/9706029.

29. J. Gegelia, Phys. Lett. B **429**, 227 (1998).

30. T. Mehen, I.W. Stewart, Phys. Lett. B **445**, 378 (1999); Phys. Rev. C **59**, 2365 (1999).

31. L.D. Landau, E.M. Lifshitz, *Quantum Mechanics: Nonrelativistic Theory* (Pergamon Press, Toronto, 1977).

32. X. Kong, F. Ravndal, Phys. Rev. D **59**, 014031 (1999); Phys. Lett. B **450**, 320 (1999).

33. See, *e.g.*, G. Miller, Phys. Rev. C **67**, 042501 (2003).

34. G. Barton, Nuovo Cimento **19**, 512 (1961).

35. B.R. Holstein, Phys. Rev. D **23**, 1618 (1981).

36. V.M. Dubovik, S.V. Zenkin, Ann. Phys. (N.Y.) **172**, 100 (1986).

37. G.B. Feldman, G.A. Crawford, J. Dubach, B.R. Holstein, Phys. Rev. C **43**, 863 (1991).

38. N. Kaiser, U.-G. Meissner, Nucl. Phys. A **499**, 699 (1989).

39. W.-Y.P. Hwang, C.-Y. Chen, Phys. Rev. C **78**, 025501 (2008).

40. E. Adelberger, W.C. Haxton, Annu. Rev. Nucl. Part. Sci. **35**, 501 (1985).

41. W. Haeberli, B.R. Holstein, in *Symmetries and Fundamental Interactions in Nuclei*, edited by W.C. Haxton, E.M. Henley (World Scientific, Singapore, 1995).

42. M.J. Ramsey-Musolf, S. Page, Annu. Rev. Nucl. Part. Sci. **56**, 1 (2006).

43. J. Lang *et al.*, Phys. Rev. Lett. **54**, 170 (1985).

44. E.G. Adelberger *et al.*, Phys. Rev. C **27**, 2833 (1983); K. Elsener *et al.*, Nucl. Phys. A **461**, 579 (1987); Phys. Rev. Lett. **52**, 1476 (1984).

45. C.A. Barnes *et al.*, Phys. Rev. Lett. **40**, 840 (1978); M. Bini *et al.*, Phys. Rev. Lett. **55**, 795 (1985); G. Ahrens *et al.*, Nucl. Phys. A **390**, 496 (1982); S.A. Page *et al.*, Phys. Rev. C **35**, 1119 (1987).

46. This point is carefully discussed in ref. [40].

47. V.A. Vesna *et al.*, Phys. Rev. C **77**, 035501 (2008).

48. C.S. Wood *et al.*, Science **275**, 1759 (1997); W.C. Haxton, C.P. Liu, M.J. Ramsey-Musolf, Phys. Rev. C **65**, 045502 (2002); W.C. Haxton, C.E. Wieman, Annu. Rev. Nucl. Part. Sci. **51**, 261 (2001).

49. C.-P. Liu, Phys. Rev. C **75**, 065501 (2007).

50. B. Desplanques, J. Missimer, Nucl. Phys. A **300**, 286 (1978); B. Desplanques, Phys. Rep. **297**, 2 (1998).

51. D.E. Driscoll, G.E. Miller, Phys. Rev. C **39**, 1951 (1989).

52. L. Girlanda, Phys. Rev. C **77**, 067001 (2008).

53. B. Desplanques, J. Benayoun, Nucl. Phys. A **458**, 689 (1986).

54. Here we follow the approach of I.B. Khriplovich, R.V. Korkin, Nucl. Phys. A **690**, 610 (2001).

55. I.B. Khriplovich, Phys. At. Nucl. **64**, 516 (2001).

56. D.O. Riska, G.E. Brown, Phys. Lett. B **38**, 193 (1972).

57. B.R. Holstein, Fiz. B **14**, 165 (2005).

58. K.A. Snover *et al.*, Phys. Rev. Lett. **41**, 145 (1978); E.D. Earle *et al.*, Nucl. Phys. A **396**, 221 (1983).

59. J.M. Potter *et al.*, Phys. Rev. Lett. **33**, 1307 (1974).

60. P.D. Evershiem *et al.*, Phys. Lett. B **256**, 11 (1991).

61. R. Balzer *et al.*, Phys. Rev. Lett. **44**, 699 (1980); S. Kystryn *et al.*, Phys. Rev. Lett. **58**, 1616 (1987).

62. D.R. Phillips, M.R. Schindler, R.P. Springer, arXiv: 0812.2073.

63. A.R. Berdoz *et al.*, Phys. Rev. C **68**, 034004 (2003).

64. J. Carlson, R. Schiavilla, V.R. Brown, B.F. Gibson, Phys. Rev. C **65**, 035502 (2002).

65. V. Yuan *et al.*, Phys. Rev. Lett. **57**, 1680 (1986).

66. L. Platter, H.-W. Hammer, U.-G. Meissner, Phys. Lett. B **607**, 254 (2005).

67. J.F. Cavaignac *et al.*, Phys. Lett. B **67**, 148 (1977).

68. In 2007-2008 University of Manitoba/TRIUMF/UNBC Winnipeg Particle and Nuclear Physics Progress Report.

69. See, *e.g.*, G.S. Danilov, Phys. Lett. B **35**, 579 (1971); C.H. Hyun *et al.*, Eur. Phys. J. A **24**, 129 (2005); D. Tadic, Phys. Rev. **174**, 1694 (1968); B. Desplanques, Nucl. Phys. A **242**, 423 (1975); K.R. Lassey, B.H.J. McKellar, Nucl. Phys. A **260**, 413 (1976); D.B. Kaplan *et al.*, Phys. Lett. B **449**, 1 (1999); M.J. Savage, Nucl. Phys. A **695**, 365 (2001); B. Desplanques, Phys. Lett. B **512**, 305 (2001); Ch.H. Hyun, T.-S. Park, D.-P. Min, Phys. Lett. B **516**, 321 (2001); R. Schiavilla, J. Carlson, M.W. Paris, Phys. Rev. C **70**, 044007 (2004); **67**, 032501 (2003).

70. V.M. Lobashov *et al.*, Nucl. Phys. A **197**, 241 (1972).

71. V.A. Knyaz'kov *et al.*, Nucl. Phys. A **417**, 209 (1984).

72. R. Sciavilla, J. Carlson, M.W. Paris, Phys. Rev. C **70**, 044007 (2004).

73. Y. Avishi, P. Grange, J. Phys. G **10**, L263 (1984).

74. D. Markoff, PhD Thesis, University of Washington (1997).

75. J. Alberi *et al.*, Can. J. Phys. **66**, 542 (1988).

76. J.M. Potter *et al.*, Phys. Rev. Lett. **33**, 1307 (1974); D.E. Nagle *et al.*, *3rd International Symposium on High Energy Physics with Polarized Beams and Polarized Targets*, AIP Conf. Proc. **51**, 224 (1978).

77. M. Simonius, in *Future Directions in Particle and Nuclear Physics at Multi-GeV Hadron Beam Facilities*, edited by D.F. Geesaman, Brookhaven National Laboratory Report BNL-52389, 147 (1993).

78. R. Schiavilla, M. Viviani, L. Girlanda, A. Kievswky, L.E. Marcucci, Phys. Rev. C **78**, 014002 (2008).

79. M. Viviani, as quoted by C. Crawford in his oral presentation at *SSP09, Taipei, Taiwan (2009)*.

80. S.R. Beans, M.J. Savage, Nucl. Phys. B **636**, 291 (2006).

Cosmic microwave background and first molecules in the early universe

Monique Signore[1,a], Denis Puy[2,b]

[1]Observatoire de Paris, LERMA, 75014 Paris, France
[2]University of Montpellier II, CNRS UMR 5024, GRAAL CC72, 34000 Montpellier, France

Received: 18 September 2008 / Published online: 12 December 2008

Abstract Besides the Hubble expansion of the universe, the main evidence in favor of the big-bang theory was the discovery, by Penzias and Wilson, of the cosmic microwave background (hereafter CMB) radiation. In 1990, the COBE satellite (Cosmic Background Explorer) revealed an accurate black-body behavior with a temperature around 2.7 K. Although the microwave background is very smooth, the COBE satellite did detect small variations—at the level of one part in 100 000—in the temperature of the CMB from place to place in the sky. These ripples are caused by acoustic oscillations in the primordial plasma. While COBE was only sensitive to long-wavelength waves, the Wilkinson Microwave Anisotropy Probe (WMAP)—with its much higher resolution—reveals that the CMB temperature variations follow the distinctive pattern predicted by cosmological theory. Moreover, the existence of the microwave background allows cosmologists to deduce the conditions present in the early stages of the big bang and, in particular, helps to account for the chemistry of the universe. This report summarizes the latest measurements and studies of the CMB with the new calculations about the formation of primordial molecules. The PLANCK mission—planned to be launched in 2009—is also presented.

Contents

ª e-mail: monique.signore@obspm.fr

ᵇ e-mail: denis.puy@graal.univ-montp2.fr

1 Elements of cosmology

Modern cosmology began in the early years of the twentieth century: first, Einstein introduced the principle of relativity in 1905 against Newton's conception of space and time [1]; next his general theory of relativity supplanted Newton's law of universal gravitation [2]. The basic theoretical structure of modern cosmology consists of a family of mathematical models derived from Einstein's gravitational theory of relativity in the 1920s by Friedmann [3], Lemaitre [4] and De Sitter [5]. Essentially, these models contain

- a description of space-time;
- a set of equations describing the action of gravity;
- a description of the bulk properties of matter.

The adoption of the "cosmological principle"—which supposes that, on large scales, the universe is homogeneous and isotropic—makes the description of space-time geometry and the form of the matter contents of the universe extremely simple. Here, let us only note that "isotropy" means that the universe looks the same in all directions and "homogeneity" means that the universe looks the same at every point.

While theoretical physicists, from Einstein's theory, started to develop the model which is nowadays called the "big-bang" model, the major steps toward the modern era were taken by observational astronomers. Hubble, as early as 1929, published his observations leading to recognition of the expansion of the universe [6]. Later, in 1965, Penzias and Wilson [7] providing a safe calibration of Bell Labs receiver and after eliminating sources of possible systematic errors, found an excess noise power equivalent to thermal radiation with a temperature of a few kelvin: this was the discovery of the CMB radiation. Its existence is strong evidence for the big-bang theory, which states that the early universe was a hot plasma of elementary particles governed by a Planckian distribution with just the temperature as a simple parameter.

First, let us emphasize that, in cosmology, quantum phenomena control the small scales of structures; gravity dominates the large scales of the structures. But let us briefly present the sequence of events that make up cosmic history in the framework of the "standard model". In the following description of the evolution of the universe, the time t refers to the total elapsed time or age of the universe, and the temperature T refers to the temperature of the radiation or often to the typical particle energy. Let us note only that there is a relation between t and T, given later in the review. Anyway, as time progresses, the universe expands and the temperature decreases. The following survey anticipates some of the explanations that we shall discuss throughout the review. A summary appears in Table 1.1.

- *The Planck epoch* ($t < 10^{-43}$ s, $T \sim 10^{19}$ GeV).

This epoch can be thought of as the beginning of time.

- *The inflationary epoch* ($t < 10^{-32}$ s, $T \sim 10^{16}$ GeV).

The physical vacuum dominates the energy, developing a repulsive field that drives the universe to enormous size. Inflation may be responsible for the "fluctuations" that led to the formation of structure. If so, one might find relics of this epoch in the fluctuations of the CMB and primordial gravitational waves.

- *The creation of radiation epoch* ($t \sim 10^{-26}$ s, $T \sim 10^{14}$ GeV).

The vacuum energy transforms itself into photons as well as particles and antiparticles of matter in equal numbers. This epoch is also called "reheating", the conversion of energy into thermal radiation. The background radiation energy we see filling the universe today originates here. It is possible that cosmic dark matter is also produced during this epoch.

- *The creation of baryonic matter epoch* (just before the electroweak transition, $T \sim 100$ GeV).

A small excess of quarks and electrons over antiquarks and positrons is generated in a process called baryogenesis. This process leaves its imprint with the presence of baryonic matter today.

- *The electroweak epoch* ($t \sim 10^{-10}$ s, $T \sim 100$ GeV).

This epoch represents the threshold of currently laboratory-tested physics. There may be cosmic relics of this epoch such as dark matter or cosmic defects.

- *The quark–hadron transition epoch* ($t \sim 10^{-4}$ s, $T \sim 200$ MeV).

The universe is supposed to make a transition from *quark soup* to hadronic matter. This transition may have left relics in the present universe such as various forms of dark matter: axions, black holes etc. This transition may have also left the matter in a clumpy state that would affect the creation of light nuclei later on.

Table 1.1 Orders of magnitude of important epochs in the history of the universe, according to the ΛCDM model

Epoch	Time	Temperature	Redshift	Physics
Planck epoch	10^{-43} s	10^{19} GeV	10^{32}	limit of spacetime
Cosmic inflation	10^{-32} s	10^{16} GeV	10^{29}	unstable vacuum
Creation of light	10^{-26} s	10^{14} GeV	10^{27}	conversion of vacuum to radiation energy
Electroweak epoch	10^{-10} s	100 GeV	10^{15}	electroweak unification
The strong epoch	10^{-4} s	200 MeV	2×10^{12}	quark–hadron transition
Weak decoupling	1 s	1 MeV	10^{12}	neutrinos decouple
e^{+}–e^{-} annihilation	5 s	0.5 MeV	5×10^{9}	electron heat dumped into photons
Nucleosynthesis	100 s	100 keV	10^{9}	nuclei formation
Spectral decoupling	1 month	500 eV	10^{6}	end of efficient photon production
Matter/radiation equality	10 000 years	10 000 K	3 300	matter dominates mass density
Last scattering epoch	0.3 My	3 000 K	1 000	universe transparent to light
Molecular epoch	15 My	1 500 K	500	formation of molecules
Dark ages	1 Gy	20 K	65	first small objects coalesce
Bright ages	2–13 Gy	3–10 K	10–30	large-scale gravitational instability
Present epoch	13 Gy	2.725 K	0	new astrophysics and physics

- *The decoupling of the weak interactions* ($t \sim 1$ s, $T \sim 1$ MeV).

It follows that the neutron-to-proton ratio is fixed and that the present universe is dominated by hydrogen. The cosmic neutrinos decoupled—as well as several other possible forms of dark matter—and have their density fixed at this time.

- *The epoch of creation of light elements* ($t \sim 100$ s, $T \sim 100$ keV).

The present abundances of helium, deuterium and lithium can provide a precise test of our understanding of this epoch.

- *The spectral decoupling epoch* ($t \sim 1$ month, $T \sim 500$ eV).

This epoch starts with the end of the efficiency of photon production. It results in the final conversion of primordial energy into the black-body spectrum of the CMB.

- *The radiation–matter transition.* The transition from the domination of radiation to that of matter appeared at about $t \sim 10\,000$ years, $T \sim 30\,000$ K or 3 eV).

- *Last scattering* ($t \sim 300\,000$ years, $T \sim 3\,000$ K).

The radiation cools enough for electrons to attach to protons—the hydrogen recombination—and the electrons almost stop interacting with the radiation. The universe becomes transparent to light. The baryonic matter decoupled from the radiation.

- At $T \sim 3\,000$ K, when helium, the first neutral atom, appeared followed by those from hydrogen, deuterium and lithium, chemistry acts up and leads to the formation of the first molecules such as HeH^{+}, H_2, HD and LiH.

In this work—and this is an original point of view—we focus on events that take place near $T \sim 3\,000$ K and which lead to CMB fluctuations and primordial chemistry as relics and observables.

This review summarizes the present knowledge of the CMB radiation, and some of the details of the generation of its possible distortions and anisotropies. The existence of other fossils—such as light nuclei—is also discussed with particular attention to the arrival of the first molecules in the universe. Earlier general reviews on CMB can be found in [8–10], and on primordial chemistry in [11].

Since the discovery of the CMB [7], measurements of CMB made in the last years have moved cosmology into a new era of precise parameter determination and the ability to probe the conditions during the early universe. The most important features were strongly indicated by the data from COBE [12], WMAP [13, 14] and many ongoing ground-based and balloon-borne observing campaigns.

Let us only note that the PLANCK satellite [15]—due to be launched in 2009—will be the first mission to map the entire CMB sky with mJy sensitivity and an angular resolution better than 10'. Past, present and future observations of the CMB radiations are summarized at the end of this article. This article is organized as follows.

- In the remainder of Sect. 1, a brief summary of the basics of the standard cosmological models is given. More details can be found in several excellent textbooks or reviews [16–18, 28, 29].
- Section 2 focuses on the various decouplings of particles in the early universe.
- Section 3 summarizes studies of the primordial nucleosynthesis and confronts the predicted primordial abundances to the observed primordial abundances.
- Section 4 reviews thermalization and processes which may generate spectral distortions of the CMB radiation.
- Section 5 describes the cosmological recombination.

- Section 6 studies primordial chemistry.
- Section 7 recalls the discovery, the main measurements of the CMB dipole and the resulting estimates of the velocity of the solar system.
- Section 8 presents the power spectrum of CMB fluctuations: temperature and polarization anisotropies.
- Section 9 discusses some of the details of the influence of primordial chemistry on CMB anisotropies (hereafter CMBA).
- In Section 10, the main results obtained by NASA satellites: COBE (Cosmic Background Explorer), WMAP (Wilkinson Microwave Anisotropy Probe) as well as the expected investigations from the future ESA PLANCK spacecraft are given.
- Section 11 gives a summary.

1.1 Space-time geometry of cosmological models

Einstein's theory involves the use of a metric tensor $g_{\mu\nu}$ that relates four-dimensional space-time intervals to a set of coordinates. Our metric signature is $(+---)$. The "cosmological principle"—which supposes that, on large scales, the universe is homogeneous and isotropic—imposes a strict symmetry on the universe and therefore a preferred time coordinate. Cosmological space-times must have the same local geometry at each point on a surface of constant time. The space-time may be expanding or contracting and different time slices differing by a scale factor $a(t)$. Then one can show that the most general space-time metric is the Robertson–Walker metric:

$$ds^2 = c^2\,dt^2 - a(t)^2\left[\frac{dr^2}{1 - kr^2} + r^2\left(d\theta^2 + \sin^2\theta\,d\phi^2\right)\right],$$
(1.1)

where t is the time coordinate, c is the speed of light, (r, θ, ϕ) are spherical polar coordinates for the spatial part of the metric and the scale factor $a(t)$ is defined so that physical lengths scale proportional to a; k is the curvature parameter with three options: $k = 0$ represents a flat universe with a Euclidean geometry on each surface of constant time; $k > 0$ signifies a closed universe, with positively curved spatial surfaces like three-dimensional versions of the surface of a sphere; $k < 0$ indicates negatively curved space sections of hyperbolic form.

From the Robertson–Walker metric, the global expansion of the spatial slices can be seen, as a function of the cosmic time, through the Hubble parameter:

$$H(t) = \frac{\dot{a}(t)}{a(t)},$$
(1.2)

where the dot means the time derivative. This model accounts naturally for Hubble's law which relates the apparent recession velocity v of a galaxy at a distance d:

$$v(t) = H_0 d,$$
(1.3)

where the zero subscript refers to the present day and H_0, the so-called Hubble constant, is the Hubble parameter $H(t)$, evaluated at the present epoch t_0.

Let us note that the Hubble constant is usually quoted as

$$H_0 = 100\,h\,\text{km}\,\text{s}^{-1}\,\text{Mpc}^{-1}.$$
(1.4)

Present observations suggest $h \sim 0.7$ [34, 35].

From the scale factor, we can define the redshift z of a source which emitted its radiation at any time:

$$\frac{\lambda_0}{\lambda} = 1 + z = \frac{a_0}{a}.$$
(1.5)

The observed wavelength λ_0 and the emitted wavelength λ are in the same ratio as the scale factor of the universe at the moment of observation a_0 and the moment of emission a. Let us only emphasize that this *effect* arises as a consequence of the light having traveled along a path through the expanding space-time.

Indeed, this wavelength shift (usually called "redshift", because the wavelengths are shifted towards large values) is not a Doppler shift. It is due to the expansion of space that stretches all wavelengths.

1.2 Dynamics of cosmological models

Einstein's field equations can be written in the following form:

$$G_{\mu\nu} = \frac{8\pi G}{c^4} T_{\mu\nu},$$
(1.6)

where the constant G is the gravitational constant of Newton; $G_{\mu\nu}$ is the Einstein tensor, which describes the action of gravity through the curvature of space-time, $T_{\mu\nu}$ is the energy-momentum tensor, which describes the bulk properties of matter. For simplicity, and because it is consistent with the cosmological principle, it is often useful to adopt the perfect fluid form for the energy tensor of cosmological matter which can be written

$$T_{\mu\nu} = \left(\rho + \frac{p}{c^2}\right)u_\mu u_\nu - \frac{p}{c^2}g_{\mu\nu},$$
(1.7)

where u_μ is the four-velocity of the fluid, ρc^2 is the energy density in the rest frame of the fluid and p is the isotropic pressure in that frame. Then the Einstein equations (1.6) simplify to the so-called Friedmann equation:

$$3\left(\frac{\dot{a}}{a}\right)^2 = 8\pi G\rho - \frac{3kc^2}{a^2} + \Lambda c^2,$$
(1.8)

to the *acceleration equation*:

$$\frac{\ddot{a}}{a} = -\frac{4\pi G\rho}{3}\left(\rho + 3\frac{p}{c^2}\right) + \frac{\Lambda c^2}{3}, \tag{1.9}$$

and to energy conservation:

$$\dot{\rho} = -3\frac{\dot{a}}{a}\left(\rho + \frac{p}{c^2}\right), \tag{1.10}$$

where Λ is called the *cosmological constant* and where the dots denote derivatives with respect to cosmic time t. These equations—which are not independent—determine the time evolution of the cosmic scale factor $a(t)$ describing the expansion or the contraction of the universe. We may note that the energy–momentum tensor $T_{\mu\nu}$ is covariantly conserved,

$$\Delta_\mu T^{\mu\nu} = 0, \tag{1.11}$$

and yields (1.10), which can be also written

$$\dot{\rho} = -3H\left(\rho + \frac{p}{c^2}\right). \tag{1.12}$$

To solve the system of equations (1.8), (1.9) and (1.10), we need to specify the equation of state, which characterizes the material contents of the universe.

Within the fluid approximation used here, we may assume that the pressure p is a single-valued function of the energy density: $p = p(\rho)$; one may define the equation-of-state parameter, ω, by

$$p = \omega \rho c^2. \tag{1.13}$$

If the universe is filled with non-relativistic matter (or cold matter), it can be described by the *dust* equation of state, with $p = 0$ and $\omega = 0$. If the universe is filled with relativistic matter (a photon gas, for instance), the equation of state of this radiation is of the form $p = \rho c^2/3$ and $\omega = 1/3$.

In the standard big-bang theory, the early universe is radiation dominated ($\omega = 1/3$); as it expands and cools, the matter becomes non-relativistic and the equation of state changes smoothly to that of dust ($\omega = 0$). For the equation of state (1.13), with a constant value of ω, (1.12) gives

$$\rho = \kappa a^{-3(1+\omega)}, \tag{1.14}$$

where κ is a constant and with the corresponding relation for energy conservation (1.12) given by

$$\dot{\rho} = -3(1+\omega)H\rho. \tag{1.15}$$

Thus, for a radiation-dominated universe (which was the case for the first several thousand years after the big bang),

$$\rho \propto a^{-4} \quad \text{(radiation domination)}, \tag{1.16}$$

and for a matter-dominated universe

$$\rho \propto a^{-3} \quad \text{(matter domination)}. \tag{1.17}$$

From the Friedmann equation (see (1.8)), we may define, at any time, the critical energy density

$$\rho_c = \frac{3H^2}{8\pi G}, \tag{1.18}$$

for which the spatial sections must be precisely flat ($k = 0$). Let us note that the critical density ρ_c is defined in terms of the expansion rate. In particular, its value today is given through the Hubble constant:

$$\rho_{c0} = \rho_c(z=0) = \frac{3H_0^2}{8\pi G} = 1.8788 \times 10^{-29}\, h^2\, \text{g cm}^{-3}. \tag{1.19}$$

We then define the density parameter by

$$\Omega = \frac{\rho}{\rho_c}. \tag{1.20}$$

Therefore the energy density of the universe is related to its local geometry:

- for $\Omega > 1$, $k > 0$, the universe is spatially closed,
- for $\Omega < 1$, $k < 0$, the universe is spatially open.

Other reduced quantities may also be used:

$$\Omega_k = -\frac{kc^2}{H^2 a^2} \quad \text{the curvature parameter,} \tag{1.21}$$

$$\Omega_\Lambda = \frac{\Lambda c^2}{3H^2} \quad \text{the cosmological constant parameter.} \tag{1.22}$$

Then the Friedmann (1.8) can be written

$$\Omega + \Omega_k + \Omega_\Lambda = 1. \tag{1.23}$$

Present-day densities in any given particle species X are quoted in units of the critical density, thus:

$$\rho_X(a_0) = \rho_{X0} = (\Omega_X \rho_c)_{a=a_0}. \tag{1.24}$$

Moreover, besides the Hubble parameter given by the relation (1.2), one may also introduce the deceleration parameter,

$$q = -\frac{\ddot{a}}{a}. \tag{1.25}$$

1.3 Some Friedmann models

Here, we do not want to discuss all the possible solutions of the Friedmann equations with a view to obtaining and classifying all universes that are homogeneous and isotropic. Let us only note that in *modern cosmology*, it is customary to specify three *observable* parameters, namely, the Hubble parameter H, the density parameter Ω and the deceleration parameter q. Usually, these *observable* parameters are estimated today and referred to by H_0, Ω_0 and q_0; in principle, they can be determined by observation. Moreover, only *dust* universes ($p = 0$ or $\omega = 0$) are considered. In this section, Sect. 1.3, we shall adopt for these particular models, instead of $a = a(t)$, the scale factor $R = R(t)$.

In general, the following cases are considered:

- static models (those which have $\dot{R} = 0$);
- empty models (with $\rho = 0$);
- the three non-empty models (with $\Lambda = 0$);
- the non-empty models (with $\Lambda \neq 0$).

Here, we shall only and briefly focus on the two last generic cases.

1.3.1 On the three non-empty models with $\Lambda = 0$

For many cosmologists and for many years, the three following models were the only models ever considered seriously:

- when the density is above the critical density:

$$\rho > \rho_c, \tag{1.26}$$

the function $R(t)$ grows from zero to a maximum value, then a collapse phase follows to zero;
- when the density is equal to the critical density, $R(t)$ is simply given by

$$R(t) = R_0 (3/2 H_0 t)^{2/3} = R_0 (t/t_0)^{2/3} \quad \text{with}$$
$$t_0 = 2/3 H_0^{-1} = 1/\sqrt{6\pi G \rho_c}; \tag{1.27}$$

- when the density is below the critical density, the function $R(t)$ grows from zero to infinity (it is easy to check from the Friedman equations that the function $R(t)$ is proportional to t when R is large);

similarly the behavior of $R(t)$ can be found when $t \to 0$ independently of the model:

$$R(t) \sim t^{2/3}. \tag{1.28}$$

1.3.2 On the non-empty models with $\Lambda \neq 0$

To specify a cosmological model, it is customary to determine with precision two *observables*: Ω_0 and q_0 for instance. The modern cosmological view of the Friedmann models is summarized in Fig. 1.1. One must note that a strong motivation for the existence of a non-zero cosmological constant comes from observations.

- Recent results from the CMBA observations (Boomerang [30], Archeops [31], WMAP [32]) show that the universe is nearly flat. Once combined with HST (Hubble Space Telescope) measurements, these results require a non-zero cosmological constant [33].

- In 1998, two independent groups—the Supernova Cosmology Project (SCP) [34] and the High-z Supernova Search (HzS) [35]—announced a spectacular result based on observations of distant type-Ia supernovae: the expansion of the universe is accelerating! Taking into account these CMBA and SNIa results and their conclusions allow us to put the sign "We are here" on Fig. 1.1.

Let us recall that the classic discussion of the physics of the cosmological constant can be found in Weinberg [36, 37]—more recent views are presented, for instance, by Straumann [38], Carroll [39], Weinberg [40].

Let us also remark that we have introduced the equation of state (1.13): $p = \omega\rho c^2$, with $\omega = 1/3$ for radiation, $\omega = 0$ for cold matter. If one considers the vacuum energy density ρ_v (or the cosmological constant Λ) such that

$$p_v = -\rho_v c^2 = -\frac{\Lambda c^2}{8\pi G}, \tag{1.29}$$

then $\omega_v = -1$.

Most cosmologists consider also for *dark energy* quintessence, a time-evolving, spatially inhomogeneous energy component. The equation of state, for quintessence, is a function of redshift: $\omega = \omega(z)$. Now, whether the dark energy is really constant as in Einstein's original version or whether it is in the form of quintessence, the question may be answered by observational plans such as the spatial mission SNAP (SuperNova Acceleration Probe); the question whether the dark energy can be understood within a fundamental theory may be *only* answered from a particle-physics viewpoint.

1.4 Inflation: a solution to the problems of the standard big-bang model

Although the Friedmann model described above seems to give a successful description of the universe, there are some fundamental problems about the initial conditions required. Three of these problems are: the "horizon problem", the "flatness problem" and the "monopole problem".

- *The horizon problem.*
The universe is uniform and isotropic on large scales. In particular, if one observes the cosmic microwave background in two regions in the sky located in directions that are widely different and which could never have been in thermal contact—see Fig. 1.2—their temperature is the same within 1 part in 10^4. How can their temperatures be so closely equal?

Fig. 1.1 Evolution of the scale factor R with the time, as predicted by Friedmann equations (from Melchiorri et al. [19])

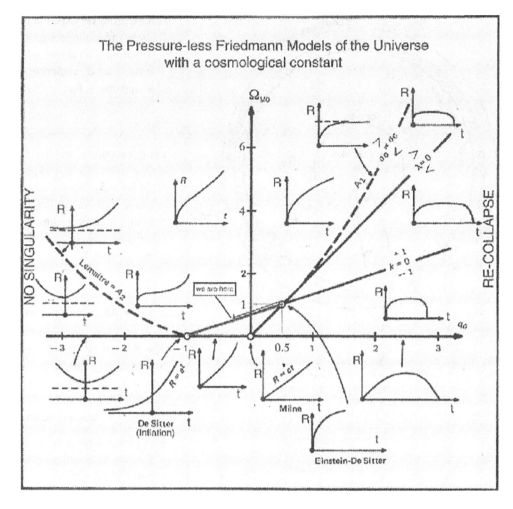

Fig. 1.1 Evolution of the scale factor R with the time, as predicted by Friedmann equations (from Melchiorri et al. [19])

- *The flatness problem.*

Another major cosmological problem is to understand why the universe is so flat, now, at $t \sim 10^{17}$ s, i.e. so close to the border between "open" or "closed". Moreover, at Planck time, where $\hbar_P$ is the reduced Planck constant,

$$t_{Planck} = \sqrt{G\hbar_P/c^5} \sim 10^{-43} \text{ s}, \tag{1.30}$$

or $T_{Planck} \sim 10^{19}$ GeV—the Planck temperature—, one can show that $\Delta\rho/\rho$ would have been $10^{-43}/10^{17} \sim 10^{-60}$. *How could Ω have been so finely tuned to unity? Or in other words, how could the universe have been so flat? How has the universe survived for $10^{60}t_{Planck}$, without recontracting or exhibiting noticeable negative curvature?* One needs a mechanism to reduce the curvature by a factor 10^{50}–10^{60}.

- *The monopole problem.*

The existence of the grand unification of forces in nature implies that superheavy magnetic monopoles have been produced when $kT \sim 10^{15}$ GeV. Calculating their relic abundance, one finds that they would have overclosed the universe unless there is some mechanism to dilute their numerical density.

Inflation [20–24] can solve the horizon, flatness and monopole problems [25] and also other problems [20]. Consider the Friedman models of the universe with $\Lambda = 0$ (see Fig. 1.1) in a situation where Λ dominated the expansion in the Friedman equation such that

$$H = \frac{1}{a}\frac{da}{dt} = \sqrt{\frac{\Lambda}{3}} = \text{constant}, \tag{1.31}$$

with the solution

$$a \sim \exp(Ht). \tag{1.32}$$

This De Sitter solution, or inflation solution, is shown in the lower left corner of the plot of Fig. 1.1.

In typical models for inflation, the phase transition of a scalar field—sometimes called the inflaton field—took place when $T \sim T_{GUT} \sim 10^{16}$ GeV at $t_{GUT} \sim 10^{-32}$ s. Inflation causes the expansion to accelerate by a factor of more than 10^{26}. This is schematically shown in Fig. 1.3. After inflation stopped, the vacuum energy of the inflaton field was transferred to ordinary particles, so a reheating of the universe took place with an enormous entropy generation.

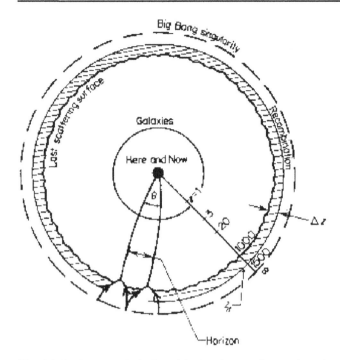

Fig. 1.2 In the cosmological model, the observer (here and now) can "understand" that the two regions—which were in causal contact—have the same temperature, within 1 part in 10^4, but not for regions which were widely separated

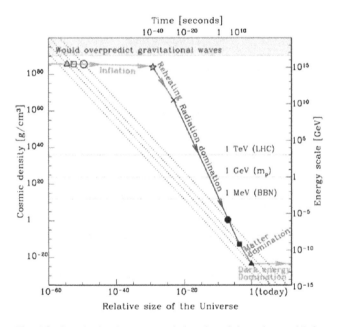

Fig. 1.3 Cosmic density versus relative size of the universe: (1) for the standard model (*dotted diagonals*), (2) for models with inflation; curves from Bock et al. [25]

The period of inflation and reheating is strongly non-adiabatic, since there is an enormous generation of entropy at reheating. After the end of inflation, the universe "restarts" in an adiabatic phase with the standard conservation of aT, and it is because the universe automatically

restarts from very special initial conditions that the horizon, flatness and monopole problems are avoided (see Fig. 1.3).

Inflation is an active subject of research for theoreticians. The most solid prediction from inflation is that the observable universe is flat:

$$\Omega = 1, \tag{1.33}$$

to high accuracy, and that fluctuations on large scales in the cosmic microwave background radiation should be nearly scale invariant. Moreover, primordial fluctuations should be adiabatic and nearly Gaussian. Finally, most of the scenarios predict a significant amount of gravitational waves, which could give rise to CMB anisotropy and influence polarization.

2 Decoupling particles in the early universe

The thermal history of the early universe is very simple. It just assumes a global isotropic and uniform universe (see Sect. 1.1). In its reduced version—no structure of any kind on scales larger than individual elementary particles—the contents of the universe are determined by *just physics*.

- A global expansion governed by general relativity.
- Particle interactions governed by the standard model of particle physics.
- Distributions of particles governed by the laws of statistical physics.

In effect, at early times ($T > 10^{12}$ K, $t < 10^{-4}$ s), the universe is filled up with an extremely dense and hot gas: the matter is completely dissociated and is in equilibrium with radiation.

A given species remains in thermal equilibrium as long as its interaction rate Γ is larger than the expansion rate H of the universe:

$$\Gamma \gg H, \tag{2.1}$$

where the Hubble parameter H sets the cosmological time scale and where the interaction rate Γ for a particle with cross section σ is typically of the form

$$\Gamma = n\sigma \langle v \rangle. \tag{2.2}$$

n is the numerical density and $\langle v \rangle$ the mean typical particle velocity. In the following, we briefly review the generic behavior of elementary particles in the early universe; we shall consider first particles which were in equilibrium with the surrounding thermal bath, then thermal relics, and finally possible relics which were never in thermal equilibrium.

2.1 Thermal equilibrium in the early universe

Let us focus on a generic population of particles ι and their antiparticles $\tilde{\iota}$, in thermal equilibrium (same temperature: $T_\iota = T_{\tilde{\iota}}$), with a neglected asymmetry between matter and antimatter ($n_\iota = n_{\tilde{\iota}}$), with fast annihilation and creation reactions ($\iota + \tilde{\iota} \rightleftharpoons a + \tilde{a}$), and with negligible chemical potentials.

For a gas of particles, we can describe the state in terms of a distribution function $f(\mathbf{p})$, where the energy E is related to the three-momentum $\mathbf{p}$ and the mass m of the species ι by

$$E^2(\mathbf{p}) = m^2 c^4 + |\mathbf{p}^2| c^2. \tag{2.3}$$

In thermal equilibrium, at a temperature T, the particles obey either Fermi–Dirac or Bose–Einstein distributions:

$$f(\mathbf{p}) = \frac{1}{\exp(E(\mathbf{p})/k_B T) \pm 1}, \tag{2.4}$$

where k_B is the Boltzmann constant, and where the plus sign is for fermions and the minus sign for bosons. The numerical density, energy density and pressure of species labeled ι are respectively given by

$$n_\iota = \frac{g_\iota}{(2\pi \hbar_P)^3} \int f(\mathbf{p}) \, \mathrm{d}^3 \mathbf{p}, \tag{2.5}$$

$$\rho_\iota = \frac{g_\iota}{(2\pi \hbar_P)^3} \int E(\mathbf{p}) f(\mathbf{p}) \, \mathrm{d}^3 \mathbf{p}, \tag{2.6}$$

$$p_\iota = \frac{g_\iota}{(2\pi \hbar_P)^3} \int \frac{|\mathbf{p}^2| c^2}{3} E(\mathbf{p}) f(\mathbf{p}) \, \mathrm{d}^3 \mathbf{p}, \tag{2.7}$$

where $\hbar_P$ is the reduced Planck constant ($\hbar_P = h_P/2\pi$), and where g_ι is the number of spin states of the particle ι—for photons $g_\gamma = 2$; for positrons or for electrons $g_e = 2$; for a massive boson Z $g_Z = 3$.

We can carry out the integrals over the distribution functions in two opposite limits: particles which are highly relativistic ($kT \gg mc^2$) or highly non-relativistic ($kT \ll mc^2$).

2.1.1 Numerical density

Assuming a system of units where the reduced Planck constant $\hbar_P$, the Boltzmann constant k_B and the speed of light c are all equal to unity, the numerical density is given by

$$n_\iota = \frac{g_\iota}{2\pi^2} T^3 \int x^2 \frac{\mathrm{d}x}{\exp(y) \pm 1}, \tag{2.8}$$

where $x = p/T$ and $y = E/T$ (the plus sign is relative to fermions when the minus sign is for bosons).

In the ultra-relativistic limit, $m_\iota \gg T$, the numerical density can be written

$$n_\iota = \frac{\zeta(3)}{\pi^2} g_\iota T^3 \quad \text{for bosons,} \tag{2.9}$$

$$n_\iota = \frac{3}{4} \frac{\zeta(3)}{\pi^2} g_\iota T^3 \quad \text{for fermions,} \tag{2.10}$$

where $\zeta(x)$ is the Riemann zeta-function.[1] To get back to the usual system of units (CGS or MKSA), the multiplicative factor $(k/\hbar_P c)^3$ must be introduced (see for instance [41]). In the non-relativistic limit, $m_\iota \gg T$, the numerical density—both for bosons and fermions—is given by

$$n_\iota = g_\iota \left(\frac{m_\iota T}{2\pi} \right)^{3/2} \exp(-m_\iota/T). \tag{2.11}$$

2.1.2 Energy density

Now let us focus on the energy density ρ_ι and on the pressure p_ι. In a system of units where $\hbar_P = k_B = c = 1$, the energy density reduces to the integral

$$\rho_\iota = \frac{g_\iota}{2\pi^2} T^4 \int \frac{x^2 y \, \mathrm{d}x}{\exp(y) \pm 1}. \tag{2.12}$$

This expression needs to be multiplied by the quantity $(k_B^4/\hbar_P^3 c^3)$ in order to restore proper units [28].

In the ultra-relativistic regime, the energy density is given by the Stefan–Boltzmann law:

$$\rho_\iota = g_\iota \frac{\pi}{30} T^4 \quad \text{for bosons,} \tag{2.13}$$

$$\rho_\iota = \frac{7}{8} g_\iota \frac{\pi^2}{30} T^4 \quad \text{for fermions,} \tag{2.14}$$

and the pressure is

$$p_\iota = \frac{\rho_\iota}{3} c^2 \quad \text{for bosons and fermions.} \tag{2.15}$$

In the non-relativistic regime—for bosons and fermions—the energy density and the pressure are respectively given by

$$\rho_\iota = m_\iota n_\iota \quad \text{and} \quad p_\iota = k_B T n_\iota \ll \rho_\iota c^2. \tag{2.16}$$

From the above expressions, several pieces of information can be extracted: relativistic particles, whether bosons or fermions, remain in approximately equal abundances in equilibrium. Once they become non-relativistic, however, their abundance plummets and becomes exponentially suppressed with respect to the relativistic species. This is simply because it becomes progressively harder for massive particle–antiparticle pairs to be produced in a plasma with $T \ll m$.

Let us note that, because the energy densities of matter and radiation scale differently, the universe was radiation

---

[1] $\zeta(3) = 1.20206$.

dominated at early epochs, while the universe is matter dominated today.

We can write the ratio of the density parameters in matter and in radiation as

$$\frac{\Omega_{\mathrm{m}}}{\Omega_{\mathrm{r}}} = \frac{\Omega_{\mathrm{m}0}}{\Omega_{\mathrm{r}0}}\frac{a}{a_0} = \frac{\Omega_{\mathrm{m}0}}{\Omega_{\mathrm{r}0}}(1+z)^{-1}, \qquad (2.17)$$

where the z redshift of the matter–radiation equality is thus

$$1 + z_{\mathrm{eq}} = \frac{\Omega_{\mathrm{m}0}}{\Omega_{\mathrm{r}0}} \sim 3300. \qquad (2.18)$$

This expression assumes that the particles that are non-relativistic today were also non-relativistic at z_{eq}; this should be an assumption, with the possible exception of massive neutrinos, which make a minor contribution to the total density.

As we already mentioned and as we shall see later—for instance for neutrinos, for photons—even decoupled, these relativistic particles will maintain a thermal distribution. This is not because they are in equilibrium, but due to cosmic expansion. In effect, during the expansion, the distribution function is transformed into a similar function but with a lower temperature, proportional to $1 + z$. We can therefore speak of the *effective temperature* of a relativistic species T_ι^{R} that freezes out at a temperature T_{f} and redshift z_{f}:

$$T_\iota^{\mathrm{R}}(z) = T_{\mathrm{f}}\frac{1+z}{1+z_{\mathrm{f}}}. \qquad (2.19)$$

A similar effect occurs for particles which are non-relativistic at decoupling, with one important difference. For non-relativistic particles, i.e. NR, the temperature is proportional to the kinetic energy, which scales as $(1 + z)^2$. We therefore have

$$T_\iota^{\mathrm{NR}}(z) = T_{\mathrm{f}}\left(\frac{1+z}{1+z_{\mathrm{f}}}\right)^2. \qquad (2.20)$$

In either case we are imagining that the species freezes out, whether relativistic or non-relativistic, and stays that way afterward; if it freezes out while relativistic and subsequently becomes non-relativistic, the distribution function will be distorted away from a thermal spectrum.

The notion of an effective temperature allows us to define a corresponding notion of an effective number of relativistic degrees of freedom, which in turn permits one to give a compact expression for the total relativistic energy density in the early universe. The effective number of relativistic degrees of freedom, g_{eff} (as far as energy is concerned), can be defined by

$$g_{\mathrm{eff}}(T) = \sum_{\iota=\mathrm{bosons}} g_\iota\left(\frac{T_\iota}{T}\right)^4 + \frac{7}{8}\sum_{\iota=\mathrm{fermions}} g_\iota\left(\frac{T_\iota}{T}\right)^4, \quad (2.21)$$

where T is the temperature of the background plasma, assumed to be in equilibrium. Then, the total energy density in all relativistic species can be written

$$\rho_{\mathrm{r}} = \frac{\pi^2}{30}g_{\mathrm{eff}}(T)T^4. \qquad (2.22)$$

2.1.3 Some important formulae of the physics of the early universe

In the framework of the Friedmann–Robertson–Walker model, (that is, for small a) the curvature term was much less important than energy-density term. Except for a possible period of inflation, the contribution due to the cosmological constant was also negligible. Therefore, the Friedmann equation, for the early universe, can be written:

$$\begin{aligned} H^2 &= \frac{8\pi G}{3}\rho_{\mathrm{r}} \\ &= \frac{8\pi G}{3}\frac{\pi^2}{30}g_{\mathrm{eff}}(T)T^4 \sim 2.76 g_{\mathrm{eff}}\frac{T^4}{m_{\mathrm{Pl}}}, \end{aligned} \qquad (2.23)$$

where m_{Pl} is the Planck mass defined by

$$m_{\mathrm{Pl}} = \sqrt{\frac{\hbar_{\mathrm{P}}c}{G}} \sim 1.221 10^{19} \text{ GeV}/c^2. \qquad (2.24)$$

Therefore, one of the most important formulae of the physics of the early universe is the expansion rate:

$$H = 1.66\sqrt{g_{\mathrm{eff}}}\frac{T^2}{m_{\mathrm{Pl}}}. \qquad (2.25)$$

Moreover, one can add that for radiation domination:

$$H = \frac{\dot{a}}{a} = \frac{1}{2t}, \qquad (2.26)$$

and the time–temperature relation becomes

$$t = 0.30\frac{m_{\mathrm{Pl}}}{\sqrt{g_{\mathrm{eff}}}T^2} \sim (1 \text{ MeV}/T)^2 \text{ s}. \qquad (2.27)$$

This is a convenient formula, valid during the important temperatures around 1 MeV when most of nucleosynthesis and neutrino decoupling occurred: when the temperature decreased from 1 MeV down to 0.1 MeV (from 10 to 1 billion kelvin); the age of the universe was in the range between one second and three minutes.

2.1.4 On the entropy of the early universe

Finally, we can also focus on the entropy of the primordial plasma. The first law of thermodynamics leads to the relation

$$Ts_\iota = \rho_\iota + p_\iota, \qquad (2.28)$$

where s_ι is the rest-frame entropy density of the gas of particles ι. In the system of units $\hbar_P = k_B = c = 1$ [28], the entropy density may be expressed by

$$s_\iota = \frac{g_\iota}{2\pi^2} T^3 \int \frac{x^2(y + x^2/3y)\,dx}{\exp(y) \pm 1}, \tag{2.29}$$

where $x = p/T$ and $y = E/T$. In the ultra-relativistic regime, we find

$$s_\iota = \frac{4}{3} \frac{\rho_\iota}{T}. \tag{2.30}$$

Therefore, the entropy density for the relativistic species is given by

$$s(T) = \frac{2\pi^2}{45} g_{\text{eff}}^S(T) T^3, \tag{2.31}$$

with

$$g_{\text{eff}}^S(T) = \sum_{\iota=\text{bosons}} g_\iota \left(\frac{T_\iota}{T}\right)^3 + \frac{7}{8} \sum_{\iota=\text{fermions}} g_\iota \left(\frac{T_\iota}{T}\right)^3. \tag{2.32}$$

Numerically, $g_{\text{eff}}(T)$ and $g_{\text{eff}}^S(T)$ are very close to each other. In the standard model, they are both of the order of 100 for $T > 300$ MeV, 10 for $1 < T < 300$ MeV and 3 for $T < 1$ MeV.

The processes that change the effective number of relativistic degrees of freedom are the QCD phase transition at $T \sim 200$ MeV, and the annihilation of electron/positron pairs at 0.5 MeV. The entropy of the gas contained in a co-volume may be expressed as

$$S = s(T)a^3 = \frac{2\pi^2}{45} g_{\text{eff}}^S(T) T^3 a^3. \tag{2.33}$$

Since entropy was only produced at a process like a first-order phase transition or an out-of-equilibrium decay, and since we suppose that the entropy production for such processes is very small compared to the total entropy, one can say that adiabatic evolution is an excellent approximation for almost the entire universe. The conservation of the comoving entropy of the primordial gas implies the following expression for the evolution of the temperature:

$$T \sim \left(\frac{1}{g_{\text{eff}}^S a^3}\right)^{1/3}. \tag{2.34}$$

Whenever g_{eff}^S is constant, one obtains the familiar result, $T \sim a^{-1}$.

2.2 Thermal relics

In general, particles did not stay in equilibrium forever; as we have already mentioned, eventually the density became

so low that interactions became infrequent, and particles froze out. Since most of the particles in our present universe belong to this category, it is important to study the features of that quenching and to compute the relic abundances of the remnant species. The decoupling has occurred either when the species ι was relativistic ($T_{\text{dec}} \gg M_\iota$) or when it was non-relativistic ($T_{\text{dec}} \ll M_\iota$). These two extreme cases are briefly reviewed in the following subsections.

2.2.1 Decoupling of relativistic particles or thermal decoupling

Let us focus on relativistic, or hot, particles ι. Their coupling to the *primordial plasma*—which was in thermal equilibrium—was achieved as long as the collision rate $\Gamma_\iota(T)$ of the ι with the other species was greater than the expansion rate $H(T)$ of the universe. As a consequence of the expansion, their temperature and their density decreased. Then, their collisions were more and more rare and less and less energetic. Finally, the relativistic particles ι could easily decouple from the thermal equilibrium, could behave as free particles, and could become a fossil radiation as soon as

$$\Gamma_\iota(T) = H(T). \tag{2.35}$$

As an example, we present the thermal decoupling of the neutrinos. Neutrino equilibrium was essentially maintained by weak interactions:

$$e^+ + e^- \rightleftharpoons \nu + \bar{\nu}, \tag{2.36}$$

where all neutrino flavors are involved. The rate of their collisions with electrons and positrons can be written

$$\Gamma_\iota(T) \sim \frac{\zeta(3)T^3}{\pi^2} \frac{G_F^2}{T^2}$$

$$\sim 4.9 \times 10^{-23} \text{ MeV} \left(\frac{T}{1 \text{ MeV}}\right)^5, \tag{2.37}$$

where the Fermi constant is $G_F = 1.16 \times 10^{-11}$ MeV^{-2}. The Hubble parameter is given by

$$H(T) = 1.66\sqrt{g_{\text{eff}}} \frac{T^2}{m_{\text{Pl}}}. \tag{2.38}$$

Comparing the collision rate with the expansion rate, we can infer a freeze-out temperature close to 1–2 MeV, below which collisions were so rare that neutrinos became a fossil radiation.

After decoupling, neutrinos moved as free particles following the general expansion. They remained in a thermal Fermi–Dirac distribution with an effective temperature which satisfied the conservation of entropy:

$$g_{\text{eff}}^S(Ta)^3 = \text{constant}. \tag{2.39}$$

After the neutrino freeze-out, electrons and positrons annihilated, dumping their entropy into the photons. Calculating before and after (e^+, e^-) decoupling, we have

$$\left(g_{\text{eff}}^S\right)_{\text{before}} = 2 + \frac{7}{8}(2+2) = \frac{11}{2}, \tag{2.40}$$

and

$$\left(g_{\text{eff}}^S\right)_{\text{after}} = 2, \tag{2.41}$$

and we find the well-known temperature difference between relic neutrinos ν and relic photons γ:

$$T_\nu = \left(\frac{4}{11}\right)^{1/3} T_\gamma. \tag{2.42}$$

It is thus easy to calculate the numerical and energy densities of the relic neutrinos. In particular, the numerical density is given from the value of the temperature:

$$n_\nu = \left(\frac{3}{4}\right)\left(\frac{4}{11}\right)n_\gamma = 112\,\text{cm}^{-3}. \tag{2.43}$$

The energy density is a function of the mass, which should be calculated numerically, but with analytical limits given from (2.13) and (2.16) for respectively $m_\nu \ll T_\nu$ and $m_\gamma \gg T_\gamma$; the latter is

$$\rho_\nu(m_\nu \gg T_\nu) = \sum_\nu m_\nu n_\nu \tag{2.44}$$

and leads to

$$\Omega_\nu h^2 = \frac{m_\nu}{93.2\,\text{eV}}. \tag{2.45}$$

Thus, a neutrino with $m_\nu \sim 10^{-2}$ eV would contribute to $\Omega_\nu \sim 2 \times 10^{-4}$. This is large enough to be interesting, without being large enough to make neutrinos dark matter! However, cosmology can provide information on the scale of the neutrino masses, complementary to the results of tritium β decay, double β decay experiments and also LEP.

For instance, analysis of the data from the anisotropies of the cosmic microwave background radiation combined with other observational and experimental results may provide upper bounds on the sum of neutrino masses. Anyway, the problem with neutrinos is that they decouple, while relativistic and hence have a comparable numerical density to that of photons. To ensure a reasonable energy density, their mass must be small.

Now we shall consider instead a species X which is non-relativistic, or cold, at the time of decoupling. It is much harder to accurately calculate the relic abundance of a cold relic than a hot one, simply because the equilibrium abundance of a non-relativistic species is changing rapidly with respect to the background plasma, and we have to be quite precise following the freeze-out process to obtain a reliable answer.

2.2.2 Decoupling of non-relativistic particles or chemical freeze-out

Let us consider some kind of particle J with antiparticle $\bar{J}$, which can annihilate each other and be pair created through the processes

$$J + \bar{J} \rightleftharpoons A + \bar{A}, \tag{2.46}$$

where A stands for any type of particle to which the J can annihilate. Moreover, we suppose that, in the early universe, the A had zero chemical potential and were in thermal equilibrium with the other light particles. The accurate calculation typically involves numerical integration of the Boltzmann equation for a network of interacting particles species; here, we cut to the chase and simply provide a reasonable approximate equation.

At $T \gg M_J$, in exact thermal equilibrium, the numerical density of J particles is $n_J^{\text{eq}}(T)$ and the rate of the above process (2.46) is the same in both directions.

When $T \ll M_J$, the creation of particles is impossible, while the annihilation is again possible. Without expansion, the particles J would disappear. With the dilution due to the expansion, annihilations will be inhibited, the collisions with the other particles will rapidly stop, and the J particles become fossils: they are the *cold relics* of the universe which pervade the intergalactic medium.

If σ_{an} is the annihilation cross section of species J at the temperature $T = M_J$, the numerical density n_J of the J evolves according to

$$\frac{dn_J}{dt} = -Hn_J - \langle\sigma_{\text{an}}v\rangle\left[n_J^2 - \left(n_J^{\text{eq}}\right)^2\right], \tag{2.47}$$

where the first term of the right-hand side of this equation expresses the dilution that comes from the Hubble expansion, the second term describes the annihilation and the third term the creation of pairs. From the time–temperature relation (2.27), one can transform the evolution equation (2.47) for n_J as a function of time into an evolution equation for n_J as a function of temperature. Then, one introduces the variable $x = M_J/T$ and normalizes n_J to the entropy density S through

$$Y_J = \frac{n_J}{S}. \tag{2.48}$$

The final equation for the normalized numerical density Y_J can be written:

$$\frac{x}{Y_J^{\text{eq}}}\frac{dY_J}{dx} = -\frac{\Gamma_{\text{an}}}{H}\left[\left(\frac{Y_J}{Y_J^{\text{eq}}}\right)^2 - 1\right], \tag{2.49}$$

where $\Gamma_{\text{an}} = n_J^{\text{eq}}\langle\sigma-\text{an}|\mathbf{v}|\rangle$; for small x, we have $Y_J = Y_J^{\text{eq}}$, since, at high T, the J particles are in thermal equilibrium. We see that the evolution is governed by the factor Γ_{an}/H, the interaction rate divided by the expansion rate. The solutions of (2.49) have to be obtained numerically to find, in particular, the temperature T_f and the corresponding value x_f of the freeze-out.

These solutions are represented on Fig. 2.1 for different values of $\langle \sigma_{an} | \mathbf{v} | \rangle$. In effect, on this figure, one can see that for $x = x_f$, Y leaves its equilibrium curve Y_{eq} and takes its actual value $Y = Y_{real}$ shown by the dashed lines. As can be seen, the value of x_f is lower for a smaller cross section σ_{an}: the more weakly coupled particles will decouple earlier and will have a higher relic abundance.

Going through the numerical analysis one finds that

- for neutral leptons of mass $M = 1$ GeV, which have essentially weak interactions: $x_f \sim 15$, $T_f \sim 70$ MeV,
- for charged leptons of mass $M = 1$ GeV, which have mainly electromagnetic interactions: $x_f \sim 34$, $T_f \sim 30$ MeV,
- for hadrons of mass $M = 1$ GeV, which have essentially strong interactions: $x_f \sim 40$, $T_f \sim 25$ MeV.

2.2.3 On dark-matter candidates

In Sect. 2.2.1, we have seen that hot relics—through (2.45)—cannot overclose the universe:

$$\Omega_\nu h^2 < 1. \qquad (2.50)$$

In Sect. 2.2.2. we have focused on cold relics—particles that were non-relativistic at time of freeze-out. Through a numerical analysis, one can find that a hypothetical neutrino with a mass close to 2–3 GeV, would have the right mass to close

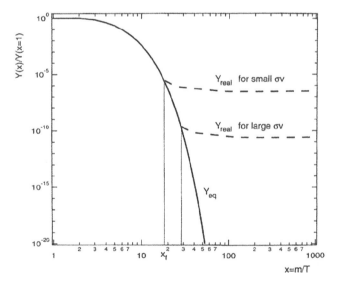

Fig. 2.1 The freeze-out of a massive particle. At a certain value $x_f = m_X / T_f$, the numerical density Y leaves an equilibrium abundance curve and takes the actual abundance showed by the *dashed lines*. For weakly interacting massive particles, x_f is order of 20 (from Bergstrom and Goobar [28])

the universe [28]. But data from the LEP accelerator have excluded neutrinos with a mass in the GeV range.

So, what could the dark matter be?

First, let us remark that the nature of cold dark-matter (hereafter CDM) particles may easily be known if they decay into radiation or some other identifiable particles. Anyway, one candidate for CDM is a weakly interacting particle (hereafter WIMP). One can show that a stable particle with a weak interaction cross section may produce a relic density of order of the critical density today.

The best example may be provided by the lightest supersymmetric partner (LSP), if it is stable and if supersymmetry is broken at the weak scale. Moreover, one could produce such particles in colliders and directly detect an hypothetic WIMP background in cryogenic detectors in underground laboratories [42].

An alternate candidate for the dark matter is provided by the axion which was not created in thermal equilibrium. For a mass range between 10^{-6} and 10^{-3} eV, the axion may produce a sizable relic density [43, 44]. Experimental efforts are currently done, in the interesting mass region, in the USA and Japan.

3 Primordial nucleosynthesis

Big-bang nucleosynthesis provides a probe of the early evolution of the universe and of its particle content. Light elements are important relics of the primordial universe. It is interesting to confront the predictions of the *standard big-bang nucleosynthesis* (SBBN) model for the abundances of deuterium, helium-4 and lithium-7 with the observations of the *primordial abundances* of these light elements.

The SBBN model has only one parameter, the baryon-to-photon ratio η. These abundances, which depend only on η, span some nine orders of magnitude. The synthesis of these elements strongly depends on the physical conditions of the early universe, when T was below 500 keV, when t was below 5 s.

But let us consider in some detail the SBBN model [45–50, 52, 53, 64–66].

3.1 Standard big-bang nucleosynthesis

The universe, after the big bang, was expanding and was filled with radiation. All wavelengths were stretched along with the expansion. At early times, the universe was dense and hot, and all particles were in equilibrium. As seen in Sect. 2, as the early universe expanded and cooled, interaction rates declined and, depending on the strength of their interactions, different particles fell out of equilibrium at different epochs. In particular, for the neutrinos, ν_e, ν_μ and ν_τ, the departure from equilibrium occurred when the universe was a few tenths of a second old and the temperature of the cosmic background radiation photons, e^+-e^- pairs, and the neutrinos was a few MeV. Now, let us focus on the neutron-to-proton ratio.

3.1.1 The neutron-to-proton ratio

As long as $T \gg 1$ MeV, neutrinos, electrons and positrons were in equilibrium through the n–p reactions [45, 46, 48]:

$$n \rightleftharpoons p + e^- + \nu_e \tag{3.1}$$

$$n + e^+ \rightleftharpoons p + \nu_e \tag{3.2}$$

$$n + \nu_e \rightleftharpoons p + e^-. \tag{3.3}$$

The neutron-to-proton ratio n/p was approximatively given by the Saha equation:

$$n/p = \exp(-\Delta m / T), \tag{3.4}$$

with $\Delta m = m_n - m_p = 1.293$ MeV. The neutrino number was supposed to be $N_\nu = 3$. When the reaction rate $\Gamma(T)$ fell below the expansion rate $H(T)$, the weak interactions were frozen out and neutrons and protons ceased to interconvert. The equilibrium abundance of neutrons at this freeze-out temperature T_f were about 1/6 the abundance of protons: $(n/p)_f \sim 1/6$. The neutrons have a finite lifetime ($\tau = 890$ s) that was somewhat larger than the age of the universe at this epoch, $t(1$ MeV$) \sim 1$ s, but they gradually decayed into protons and leptons. Soon thereafter, however, the universe reached a temperature somewhat below 100 keV, and big-bang nucleosynthesis began. At that point, the n/p ratio had approximately the following value:

$$(n/p)_{nuc} \sim \frac{1}{7}. \tag{3.5}$$

The universe was about 100 second old.

3.1.2 Helium synthesis

While neutrons and protons were interconverting, they were also colliding among themselves, creating deuterons:

$$n + p \rightarrow D + \gamma. \tag{3.6}$$

At early times, the deuteron was photodissociated before it could capture a neutron, or a proton, or another deuteron:

$$D + \gamma \rightarrow n + p \tag{3.7}$$

to build the heavier nuclides.

This famous *bottleneck* to big-bang nucleosynthesis persisted until $T \sim 0.1$ MeV. Below this temperature, photodesintegration became inefficient; then, nuclei more complex than deuterium were built through the following reactions (see Fig. 3.1):

$$n + D \rightarrow {}^3H, \tag{3.8}$$

$$p + {}^3H \rightarrow {}^4He, \tag{3.9}$$

$$p + D \rightarrow {}^3He, \tag{3.10}$$

$$n + {}^3He \rightarrow {}^4He, \tag{3.11}$$

$$D + D \rightarrow {}^4He. \tag{3.12}$$

Since there are no stable mass-5 nuclides, a *gap* appears at mass-5. The few reactions that bridged the mass-5 gap led mainly to mass-7: ${}^7Li$ and ${}^7Be$.

Finally, heavier elements were produced in truly negligible quantities, since there was another gap at mass-8.

Fig. 3.1 Network of nuclear reactions that determines the yields in standard big-bang nucleosynthesis (from Nollett and Burles [63])

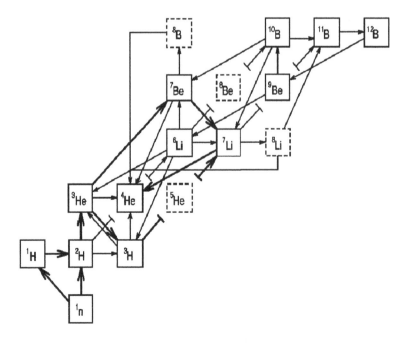

In fact, there were only some traces of beryllium-9 and boron-11. The primordial nuclear reactor was short-lived.

In the framework of the SBBN model, as the temperature dropped below $T < 30$ keV, when the universe was close to 20 min old, Coulomb barriers suppressed all nuclear reactions. Afterwards, until the first stars formed, no new nuclides were created.

Before presenting the predictions of the SBBN model, let us note that the dominant product of big-bang nucleosynthesis was ^{4}He, and its abundance was very sensitive to the $(n/p)_{nuc}$ ratio. Introducing $n_{tot} = [n + p]_{nuc}$, the abundance of helium ^{4}He that forms is

$$Y(^4He) = \frac{n_{^4He}}{n_{tot}} = \frac{2(n/p)_{nuc}}{1 + (n/p)_{nuc}} \sim 0.25, \qquad (3.13)$$

i.e. an abundance of ^{4}He close to 25% by mass. Lesser amounts of the other light elements were produced: D at the level of about 10^{-5} by number and ^{7}Li at the level of 10^{-10} by number.

3.1.3 Predicted abundances of light elements

In the SBBN model with $N_\nu = 3$, the only free parameter is the density of baryons, Ω_b or η, the ratio of the number of baryons n_b to the number of photons n_γ:

$$\eta = \frac{n_b}{n_\gamma}. \qquad (3.14)$$

As will be seen from SBBN, and as is confirmed by a variety of astrophysical data, η is very small, so it is convenient to also introduce $\eta_{10} = 10^{10}\eta$ and to use it as one of the parameters for BBN.

Having dealt with the breaking of weak equilibrium between neutrons and protons, one has to consider the onset of nuclear reactions which build up the light nuclei. This has been studied by numerical solution of the complete nuclear reaction network.

The initial Wagoner [45] numerical code—which has been significantly improved and upgraded by Kavano [61]—is again available for public download [62].

Later, Sarkar [49] discussed the elemental abundances with a semi-analytical method estimating reaction rates and uncertainties in primordial nucleosynthesis. In their poster for the APS centennial meeting, Burles et al. [51] showed that after five minutes—see Fig. 3.2—most of the neutrons were in helium-4 nuclei, most protons were free, while much smaller amounts of deuterium, helium-3 and lithium-7 were synthesized. This elemental composition was unchanged until the formation of the first stars.

The yields of primordial nucleosynthesis are shown as a function of baryon density $\Omega_b h^2$ or the baryon-to-photon ratio η in Fig. 3.2 [65].

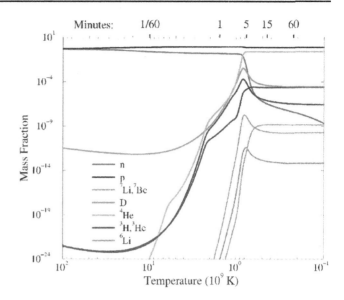

Fig. 3.2 Time–temperature evolution of the primordial nucleus in standard cosmological model (from Burles and Nollett [51]). Let us note that at very high temperatures the number of protons should be equal to the number of neutrons

The four curves of Fig. 3.3 represent the abundance ratio predicted by SBBN: the top curve is the helium-4 mass as a fraction of the mass of all baryons, while the three lower curves are the number fractions of deuterium, helium-3 an lithium-7 with regard to hydrogen.

In the standard model with $N_\nu = 3$, the only free parameter is the density of baryons, which sets the rates of the strong reactions.

Thus, any abundance measurement determines $\Omega_b h^2$ or η, while additional measurements *overconstrain* the theory and thereby provide a consistency check. SBBN has historically been the prime means of determining the cosmic baryon density.

For instance, in Fig. 3.3, the five boxes represent some of the various measurements of primordial helium-4, deuterium, helium-3 and lithium-7, but deuterium plays the historical role of first-rate *baryometer* since the vertical band that covers the D/H data determines the baryon-to-photon ratio: $\eta_{10} = 5.9 \pm 0.5$. This value corresponds to a cosmological baryon density of

$$\Omega_b h^2 = 0.0214 \pm 0.0020. \qquad (3.15)$$

As Kirkman et al. [65] remarked, it is expected that all the data boxes of Fig. 3.3 should overlap the vertical band that covers the D/H data. They do not, probably because of systematic errors. But since 2006, with the new CMB anisotropy measurements—which will be described in Sects. 8 and 10—CMB may be a much better *baryometer* than SBBN.

Recently, the team of the satellite WMAP (Wilkinson Microwave Anisotropy Probe) released their analysis after

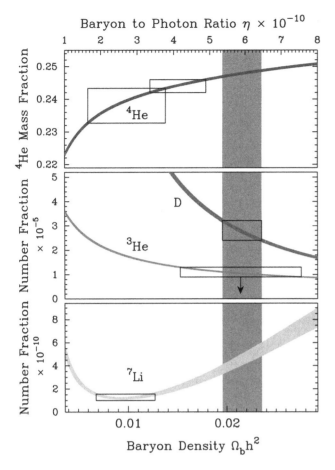

Fig. 3.3 Comparison of predicted and measured abundances of D, ^{3}He, ^{7}Li and Y_p as a function of the baryon-density parameter $\Omega_b h^2$ or the baryon-to-photon ratio η. The four curves show the predicted abundance by the SBBN model. The five boxes represent some of the measurements of primordial abundances (the larger ^{4}He box from Olive et al. [71], the smaller ^{4}He box from Izotov and Thuan [69], the deuterium box from Kirkman et al. [65], the ^{3}He box from Bania et al. [83], the ^{7}Li box from Ryan et al. [79])

three-year data taking of microwave background anisotropy measurements and gave the following value for the baryon density [66]:

$$\Omega_b h^2 = 0.0223 \pm 0.0007, \qquad (3.16)$$

which corresponds, for the baryon-to-photon ratio $\eta_{10} = (273.9 \pm 0.3)\Omega_b\, h^2$, to the approximate value

$$\eta_{10} \sim 6.11 \pm 0.20, \qquad (3.17)$$

with $\Omega_b h^2$ fixed by the CMB radiation, and precision comparisons to the most recent observations can now be attempted (see later below, the value given by the five-year WMAP data in Sect. 10).

3.2 Observations of primordial abundances

The primordial abundances of the light elements are not measured easily and simultaneously. The main difficulties come from systematic uncertainties in inferring abundances from observations and in modeling their chemical evolution since the big bang. As a propaedeutic, one describes a generic program which may be done for each of the nuclei.

3.2.1 A generic program for obtaining primordial abundances

- *Observations.* The program begins with observations of, in general, some atomic or molecular spectral line. Moreover, in the case of deuterium, for instance, one may observe one isotope in the presence of another one, which is 10 000 or 100 000 times more abundant. Finally, the various isotopes may be observed in different astrophysical sites, because of their different properties.

- *Conversion of observed lines into abundances.* In general, each isotope is detected by measuring an atomic or a molecular spectral line. It is a non-trivial task to obtain, from this observed line, an abundance of the isotope, which is the ratio of the number density of this isotope to the numerical density of hydrogen.

- *Corrections for chemical evolution.* Except for deuterium, all isotopes have potential sources which could have increased their abundances since the epoch of primordial nucleosynthesis. All, except helium-4, have significant sinks, which could have decreased their abundances since the epoch of primordial nucleosynthesis. Therefore to arrive at a *primordial abundance*, one must have: (1) a theoretical understanding of all the mechanisms of production and destruction of the isotope, and (2) several observations of different objects in order that all the free parameters, introduced in the previous steps, can be determined.

- *The primordial abundance for each isotope.* Finally, one arrives at a value for each abundance with the required precision. This required precision had varied with time, and it is in general just at the level of, or slightly beyond, current capabilities.

3.2.2 Primordial abundance of helium-4

- *Observations.*

  Helium-4 can be observed in galactic and extragalactic HII regions—regions of hot and ionized H gas—using either optical or radio recombination lines. However, the best determinations come from observations of (HeII, HeI) recombination lines in metal-poor extragalactic HII regions.

- *Conversion of observed lines into abundances.*
  Deriving an abundance from the observed lines should be

straightforward. However, corrections must be applied to compensate for excitation effects.

• *Corrections for chemical evolution.*

In stars, most of the hydrogen is converted into helium-4, which is converted into heavier elements. Then an excess of helium-4 can return into the interstellar medium. Therefore, one must account for a possible helium enrichment.

• *Primordial abundance for helium-4.*

There is a lot of independent recent and less recent observations and analyses leading to some values of Y_p. For illustration, Fig. 3.4 gives a compilation of the first measurements of the helium-4 mass fraction as a function of the oxygen abundance, which is an indicator of stellar processing.

First, there have been two, largely independent, estimates of Y_p, based on analyses of large data sets of low-metallicity, extragalactic HII regions: the Izotov and Thuan estimates [68, 69] and the Olive and Steigman [70–72]. More recently, Izotov and Thuan have expanded their original data set and attempted to account for uncertainties [73, 74]. Note also other work done on estimates of Y_p by Gruenwald et al. [75], Fukugita and Kawasaki [76], and Peimbert et al. [77]. Finally, recently, Steigman [67] has described and analyzed all these observational data and has chosen for the primordial helium-4 abundance

$$Y_p = 0.240 \pm 0.006. \qquad (3.18)$$

3.2.3 Primordial abundance of lithium-7

• *Observations.*

Lithium absorption lines can be relatively easily observed and measured in stars. There is a wealth of data on the abundances of lithium-7 in hot population-II halo stars, since the

first measurements of Spite and Spite [78]. The most relevant data come from studies of the very metal-poor stars in the halo of the galaxy or in globular clusters.

• *Conversion of observed lines in abundances.*

The observed line gives the stellar photospheric abundance via a standard stellar atmospheric technique. For stars with a surface temperature $T > 5\,500$ K and a metallicity less than about 1/20th the solar metallicity, the abundances practically show no dispersion. The famous *lithium plateau* for such stars is shown in Fig. 3.5.

• *Corrections for chemical evolution.*

The problem arises in relating the observed abundance to the cosmic abundance, i.e. the abundance of the star when it formed. There is a possibility that lithium has been depleted in these stars, though the lack of dispersion in the lithium data—at least in the hotter stars as seen in Fig. 3.5—limits the amount of depletion. Moreover, since the big bang, lithium-7 has been also produced through spallation by cosmic rays and also synthesized in type-II supernovae, via the neutrino process. Ryan et al. [79, 80] have undertaken some observations—in particular, studies of possible correlations between Li and Fe—to constrain the evolution of lithium-7.

The primordial abundance of lithium-7.

Ryan et al. [80], from data from studies of halo stars, find

$$\left(\frac{\mathrm{Li}^7}{\mathrm{H}}\right)_p = 1.23^{+0.340}_{-0.16} \times 10^{-10}, \qquad (3.19)$$

in conflict with the result derived by Bonifacio et al. [81, 82], from a sample of globular cluster stars:

$$\left(\frac{\mathrm{Li}^7}{\mathrm{H}}\right)_p = 2.19^{+0.460}_{-0.38} \times 10^{-10}. \qquad (3.20)$$

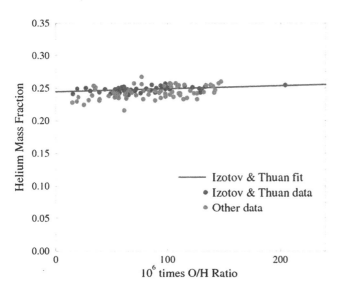

Fig. 3.4 Compilation of measurements of ^{4}He mass fraction versus O/H (indicators of stellar processing) from HII regions (large clouds of ionized hydrogen) (from Burles and Nollett [51])

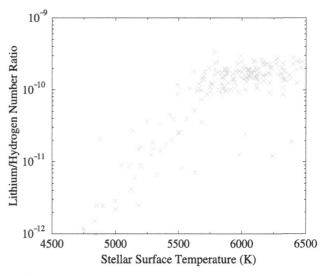

Fig. 3.5 Measurements of the ^{7}Li/H ratio as a function of stellar surface temperatures from the atmosphere of population-II stars in the halo of our galaxy. The *lithium plateau* for the stars with higher surface temperature is evident (from Burles and Nollett [51])

3.2.4 Primordial abundance of deuterium

The determination of the primordial abundance of deuterium presents certain difficulties: all deuterium is primordial, but some of this deuterium has been destroyed.

There are many measurements of the D/H ratio in the interstellar medium. But we do not know the history of the galaxy well enough to reconstruct the primordial D from galactic observations, while it is possible to detect D in high-redshift low-metallicity quasar absorption line systems. In fact, the primordial deuterium abundance has been measured in high-z hydrogen clouds, which are *seen* by their distinctive Lyman-α features in the spectra of several quasars; see [65] and references therein.

Figure 3.6 shows this feature in the spectrum of the quasar QSO 1937-1009. In his recent review, Steigman [67] describes, analyses and discusses the reasonably firm deuterium detections. Finally, he adopts for the primordial abundance deuterium

$$\left(\frac{D}{H}\right)_p = 2.68^{+0.270}_{-0.25} \times 10^{-5}. \tag{3.21}$$

3.2.5 Primordial abundance of helium-3

Finally, for the primordial abundance of helium-3, one can adopt the upper limit measured by Bania et al. [83]:

$$\left(\frac{He}{H}\right)_p < 1.1 \pm 0.2 \times 10^{-5}. \tag{3.22}$$

But helium-3 can be better seen as a probe of the galactic chemical evolution than a cosmological probe.

3.2.6 Conclusions on observations of primordial abundances

All the above primordial abundances of helium-4, lithium-7 and deuterium can be summarized in Fig. 3.7, due to Steigman [67]. This figure shows the predicted values of η_{10} corresponding to the various primordial abundances given in previous subsections and the value η—CMB/LSS inferred from the three-year WMAP data [14], and the large-scale structure data [84]:

$$\eta_{CMB/LSS} = 6.11 \pm 0.20. \tag{3.23}$$

If the CMB/LSS data are taken to be the best baryometer, there is agreement on the predictions of the SBBN model

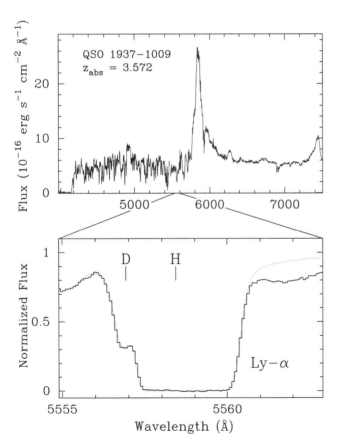

Fig. 3.6 D/H measurements, through the spectrum of the quasar QSO 1937-1009, which show the Lyman-α absorption feature from a cloud at $z = 3.572$ (from Burles and Nollett [51])

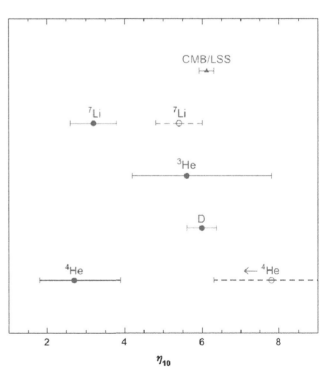

Fig. 3.7 The SBBN-predicted values of η_{10} (from Steigman [67]), their 1σ uncertainties (*red filled circles*), corresponding to the primordial abundances adopted by Steigman [67] and to the values CMB/LSS inferred from WMAP [14] and LSS [84] data (*blue triangle*). The *open circles* and *dashed lines* correspond to the alternate abundances proposed for ^{4}He and ^{7}Li by Steigman [67]

for the abundance of D and the observations of D but not for primordial ^{4}He and ^{7}Li. This disagreement between predictions and observations for ^{4}He and ^{7}Li is, as Kirkman et al. [65] and Steigman [67] said, certainly due to systematic uncertainties. More observations and more theoretical work needs to be done.

3.3 On non-standard BBN models

The above bounds on η were obtained in the framework of the standard BBN model, which assumes one and only one parameter η. But variant models of BBN exist. Below, we briefly discuss some deviations from SBBN.

3.3.1 On non-standard BBN models of Steigman's Group

The first departure from standard BBN is to assume that

(i) the number of neutrinos species is $N_\nu = 3 + \Delta N_\nu$;
(ii) there is a significant neutrino asymmetry parameter: $\xi_e = \mu_e / k_B T$, where μ_e is the chemical potential of ν_e. In particular, Kneller and Steigman [67, 85] have studied a BBN model with the three parameters η, ΔN_ν and ξ_e and showed that the ranges of the BBN parameters permitted by the data are small. They also show that D and ^{4}He abundances can constrain some classes of non-standard physics.

3.3.2 On inhomogeneous BBN models

There are possible variants of primordial nucleosynthesis with more parameters than the only one, η, of the SBBN model. The most motivated departure from standard BBN is the possibility that nucleosynthesis occurred in an inhomogeneous medium.

In the framework of these *inhomogeneous big-bang nucleosynthesis* (or IBBN) models, the formation of baryon inhomogeneous regions took place, for example during a first-order quark–hadron phase transition [54–56] or during the electroweak phase transition [57–59].

Moreover, inhomogeneities may have a particular symmetry. Lara et al. [60], who review various IBBN models, show that although the parameter space is more and more limited by more and more observations, interesting regions remain and IBBN continues to be a possibility for the early universe.

3.4 Conclusions

In the comparison of the abundance data with the theoretical expectations, we have noted the rather unsatisfactory state of the observational situation today. Whereas there has been some concerted effort in recent years towards precise abundance determinations, the quoted numbers, in particular for the ^{4}He and ^{7}Li abundances, are still plagued by uncertain systematic errors. Moreover, CMB is a better baryometer than D.

Our understanding of the CMB has been revolutionized by the accurate and consistent database provided first by the COBE mission and then by the WMAP mission. A similar revolution is overdue for primordial nucleosynthesis.

4 The cosmic microwave background spectrum

Beyond any doubt, the CMB represents the first and best studied *cosmic relic*. The CMB provides a strong observational foundation for the standard cosmological scenario, the big-bang theory.

The success of primordial nucleosynthesis calculations requires a cosmic background radiation (CBR) characterized by a temperature of $k_B T \sim 1$ MeV at a redshift of $z \sim 10^9$. In their pioneering works, Gamow, Alpher and Herman [86–88] realized this requirement and predicted the existence of a faint residual relic of the primordial radiation, with a present temperature of a few kelvin.

The observed CMB is interpreted as the current manifestation of the hypothesized CBR. In 1964, the CMB was experimentally discovered by Penzias and Wilson [7]. The theoretical explanation of Dicke's Princeton team was published in an accompanying paper [89]. Then, the following decades saw huge advances in the measurement of the CMB radiation, with COBE's definitive discovery of anisotropies and measurement of a near perfect black-body spectrum.

The small deviations from isotropy—measured in particular by COBE and later by WMAP—have told us and will continue to tell us a great deal about the inhomogeneities in our universe. The small deviations from a black-body spectrum can also tell us about the energetics in our universe. Such deviations have already been discovered in the direction of clusters of galaxies. But the mean CMB radiation spectrum is, so far, indistinguishable from a black-body spectrum at 2.725 K.

This chapter gives an introduction to the observed spectrum of the CMB radiation, reviews observations of this CMB spectrum and discusses what can be learned about it. Earlier reviews and proceedings of conferences on the CMB spectrum can be found in [90–103].

4.1 On the thermal nature of the CMB spectrum

In the framework of the big-bang theory, extrapolating the expansion back to early epochs, the universe would have been hot and dense, and the universe would have been expanding rapidly in order to grow as large as it now can be observed to be. When the matter in the universe is hot and dense, the time to reach equilibrium is short. Thus, matter rapidly approaches thermal equilibrium and photons should have a black-body spectrum at early times.

As already said, wavelengths are stretched with the cosmic expansion, while the frequencies scale inversely due to the same effect. The effect of cosmic expansion on the initial black-body spectrum is to retain its black-body nature, but just at lower and lower temperatures or lower and lower redshifts, since $T \sim 1 + z$.

Thus, as a first approximation we expect today's photons to have a black-body spectrum. The fact that the CMB radiation has nearly a black-body spectrum is strong evidence in support of the big-bang hypothesis. However, as discussed below, one can estimate that for each atom, the universe includes more than 10^9 photons. With so many more CMB photons than protons or electrons, matter cannot significantly distort the spectrum of the CMB radiation. The observed CMB spectrum, being so close to a black-body spectrum, should come as no surprise.

Now, let us recall some properties of a Planckian spectrum. In particular, the photon distribution $\eta_\gamma^{BB}(\nu)$—the photon density per unit of frequency interval—for a black-body spectrum is given by

$$\eta_\gamma^{BB}(\nu) = \frac{1}{\exp(h_P \nu / k_B T) - 1}. \tag{4.1}$$

With the observed present CMB temperature of 2.725 K, we can calculate the number of photons in the CMB (see (2.5)):

$$n_\gamma = \frac{2\zeta(3)}{\pi^2} T_\gamma^3 = 413 \left(\frac{T_{CMB}}{2.725\ K}\right)^3\ cm^{-3}. \tag{4.2}$$

We can also compute the present energy density in CMB photons:

$$\rho_\gamma = \frac{\pi^2}{15} T_\gamma^4 \sim 4.68 \times 10^{-34}\ g\,cm^{-3}, \tag{4.3}$$

and the entropy of CMB radiation is given by

$$s_\gamma = \frac{4}{3} \frac{\rho_\gamma}{T_\gamma} = \frac{4\pi^2}{45} T_\gamma^3, \tag{4.4}$$

while the ratio s_γ / n_γ is given by

$$\frac{s_\gamma}{n_\gamma} = \frac{2\pi^4}{45\zeta(3)} \sim 3.6. \tag{4.5}$$

Since—during the cosmic expansion—the temperature scales as $T_0 = T_i(1 + z_i)$, we can calculate the photon numerical density, $n_{\gamma i} = (1 + z_i)^3 n_\gamma$, and energy density, $\rho_{\gamma i} = (1 + z_i)^4 \rho_\gamma$, for any epoch i with redshift z_i. Moreover, as we have seen in Sect. 1, we can introduce the dimensionless Ω_γ through the relation: $\rho_\gamma = \rho_c \Omega_\gamma$.

The fraction of total energy in the universe in the form of relic radiation ($T_\gamma = 2.725$ K) is thus

$$\Omega_\gamma h^2 \sim 4 \times 10^{-5}. \tag{4.6}$$

The energy density at present is thus dominated by matter, since $\Omega_m h^2 \sim 0.127$. But let us also remark that $n_b/n_\gamma \sim 10^{-10}$.

One of the problems of modern cosmology is to understand why the present baryon number is so small. Another way to describe the puzzle for cosmologists could be formulated thus:

How did the universe evolve from early times, in which there were equal numbers of baryons and antibaryons, to the present universe in which there is a precisely measured—through BBN and CMB—"baryon asymmetry of the universe"?

All the solutions to this puzzle provide examples of the interplay between particle physics and cosmology. In effect, many theories beyond the standard model of particle physics when considered in the context of the expanding universe could lead to mechanisms which could explain the "baryon asymmetry of the universe". Moreover, future experiments, such as the Large Hadron Collider (hereafter LHC) and the International Linear Collider (hereafter ILC), will refute or confirm these mechanisms.

If the cosmic microwave background radiation decoupled from matter at a redshift of about 1 000, the scale factor of the universe was about 1 000 times smaller than its current size, and the original wavelength of the radiation was 1 000 times smaller than that observed today.

In Sect. 5, we shall describe the *recombination epoch* and we shall make precise the redshift of the matter–radiation decoupling.

In order to fully understand the various observations of the CMB radiation, a few words of a technical nature are needed. The spectral brightness I_ν is defined as the incident energy per unit area, per unit solid angle, per unit frequency, per unit time. It may be written

$$I_\nu = \frac{2h_P \nu^3}{c^2} \eta_\nu^{BB}, \tag{4.7}$$

where η_ν^{BB} is given by (4.1). Therefore, in natural units I_ν (in W Hz^{-1} m^{-2} sr^{-1}) we have

$$I_\nu = \frac{2h_P \nu^3}{c^2} \frac{1}{\exp(h_P \nu / k_B T) - 1}. \tag{4.8}$$

The high-frequency ($h_P \nu \gg k_B T$) limit of the Planck spectrum is known as Wien's law:

$$I_\nu^W = \frac{2h_P \nu^3}{c^2} \exp(-h_P \nu / k_B T), \tag{4.9}$$

while the low-frequency ($h_P \nu \ll k_B T$) limit of the Planck spectrum is known as the Rayleigh–Jeans law:

$$I_\nu^{RJ} = \frac{2h_P^2 k_B T}{c^2}. \qquad (4.10)$$

4.2 On observations of the CMB spectrum

Since 1946–1948, a firm prediction for the existence of a universal black-body radiation was available in the scientific literature. In 1950, Gamow [104] rederived the value of the CMB temperature in a paper published in Physics Today, giving a value of 3 K. The computation was repeated in 1953, for a Danish journal [105], leading to 7 K and in 1956, for Vista in Astronomy [106], to a value of 6 K.

Alpher and Herman also published these predictions several times [107–109]. Later on, Alpher and Herman [110] have underlined that the first prediction of CMB is in their 1948 paper [88] and not in the previous papers by Gamow [86, 87].

We think that the first prediction has to be attributed to Gamow. Let us also note that Gamow was never strongly interested in establishing the priority of his CMB prediction, because he was convinced that CMB was well below the possibility of any detection.

After this long period of prediction, we can easily distinguish three epochs:

(i) the epoch of the discovery;
(ii) the epoch of the confirmation of the Planckian character of the spectrum;
(iii) the epoch of the search for spectral distortions.

4.2.1 From the discovery until COBE

By the 1960s, Dicke and his colleagues, at Princeton University in New Jersey (USA), had begun to build the hardware required to search for the CMB radiation. Simultaneously, Penzias and Wilson of the Bell Telephone Laboratories—involved in the NASA Project ECHO—were calibrating microwave detectors in order to make astronomical observations in the centimeter waveband. Before the Princeton Group could complete a measurement of $T_{\gamma 0}$, Penzias and Wilson had observed a weak background signal at the radio wavelength of 7.35 cm in the large horn antenna of Holmdel, also in New Jersey. Their measurements did indicate a black-body radiation at 3.5 K:

$$T_{\gamma 0}(7.35 \text{ cm}) = 3.5 \text{ K} \pm 1 \text{ K}. \qquad (4.11)$$

This observation was published in 1965 [7] under the modest title *A Measurement of Excess Antenna Temperature at 4080 MHz* along with the paper by Dicke, Peebles, Roll and Wilkinson [89] appearing as a companion article to explain the fundamental significance of this measurement.

This detection of the CMB radiation was recognized as the most spectacular evidence supporting the big-bang theory after Hubble's discovery of the expansion of the universe. But it is also very important to emphasize that Penzias and Wilson had only measured a radiation flux at a single wavelength.

One must note that the CMB radiation was first seen via its effect on the interstellar CN radical by, for instance, Adams [111], McKellar [112] in 1941, Herzberg [113] in 1950 and others. For an example, McKellar [112], from data due to Adams [111] relative to the absorption of CN in an interstellar molecular cloud derived in 1941 the excitation temperature of the rotational states to be around 2.3 K, with no obvious source of excitation.

But one must also remark that the significance of these data was not realized until after 1965 by, in particular, Thaddeus [114] in 1972, Kaiser and Wright [115] in 1990, Roth et al. [116] in 1993 and others.

Let us only note that these interstellar CN measurements at wavelengths 2.64 and 1.32 mm provided cosmologists with a way to determine $T_{0\gamma}$ that was completely independent of ground-based or rocket-borne techniques.

Generally speaking, the measurements of the CMB spectrum could be divided in three epochs. The first set of measurements, carried out before 1975, were essentially directed towards testing the existence and the nature of CMB, such as the millimetric observations done by the Firenze Group of Melchiorri [19, 117].

The second set of measurements, covering the period from 1975 up to 1990, observing primarily the low-frequency (Rayleigh–Jeans) region of the spectrum, was dominated by the results of the so-called *White-Mountain Collaboration* (Berkeley–Bologna–Milano–Padua); all these results are presented in many reviews [19, 118].

The last set of observations were from rocket-borne experiments and from the COBE satellite. Most of these data are shown on Fig. 4.1.

4.2.2 The CMB spectrum from FIRAS aboard COBE

Before describing these COBE data on the CMB spectrum, let us give some details. The three experiments aboard the COBE—the Cosmic Background Explorer—satellite are the Far-InfraRed Absolute Spectrophotometer (60–630 GHz, FIRAS), the Differential Microwave Radiometers (30–90 GHz, DMR), and the Diffuse InfraRed Background Experiment [2.2; 2.4] μm, DIRBE). Mather, Principal Investigator of FIRAS and Smoot, Principal Investigator of DMR, were awarded the 2006 Nobel Price in physics.

The FIRAS experiment has given an absolutely beautiful measurement of the flux from the sky between 2 and 20 cm^{-1} (i.e. 6–630 GHz). The intensity of the background sky radiation is consistent with a black body at [119]:

$$T_{COBE} = 2.728 \pm 0.002 \text{ K}. \qquad (4.12)$$

Fig. 4.1 Collection of selected
data, until COBE, relative to the
CMB spectrum

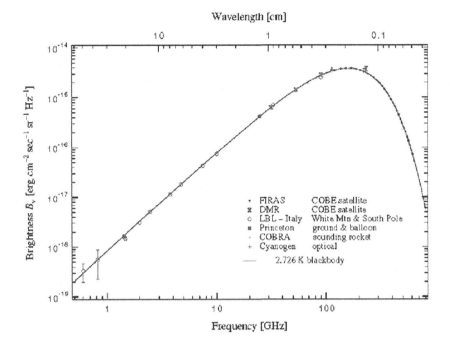

When the FIRAS investigators released their results, they claimed that deviations from this black-body spectrum, at the spectral resolution of the instrument, are less than 1% of the peak brightness. The measured spectrum by FIRAS is shown in Fig. 4.2.

The peak of the CMB is near $\nu \sim 5.5$ cm^{-1} $\sim$165 GHz. The intensity measured there is roughly 385 MJy sr^{-1}. Figure 4.3 shows the deviations from a black-body spectrum published by Mather et al. [120].

Note that measurements of the CMB spectrum, at the present level of sensitivity, face significant problems of foreground contamination. Figure 4.4 shows the CMB anisotropy (see Sect. 8) and the various major components of the microwave sky [118, 121]. Emission due to galactic synchrotron radiation from energetic electrons in interstellar magnetic fields, and due to bremsstrahlung from thermal

electrons in ionized HII regions dominates at very low frequencies. At high frequencies, the dominant source is cold interstellar dust.

The decrease of the galactic electron emission with frequency along with the increasing dust emission combine to create a broad window throughout the microwave region where the cosmic background dominates. However, since we cannot expect to observe the CMB radiation outside of the galaxy, this is a fundamental limitation. Many of the results given by the FIRAS team include significant corrections for these contaminations.

On the other hand, while the bolometers of FIRAS revolutionized the field, making obsolete most other high-frequency measurements of the CMB radiation spectrum, they did not observe the low-frequency range of the spectrum. Therefore, it is important that some groups are pursu-

Fig. 4.2 The first FIRAS result
(Mather et al. 1990). The small
squares show measurements
with a conservative error
estimate of 1%. The *full line* is a
fit to the black-body form

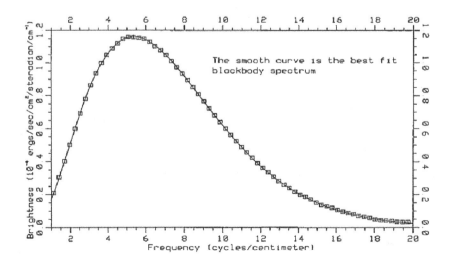

Fig. 4.3 Results from FIRAS published by Mather et al. in 1994. Data points show deviations from the black-body spectrum with the temperature 2.727 K (the final result for the temperature is 2.725 ± 0.002 K). The largest deviations were 0.03 percent of the maximum radiation; see Mather et al. [120]

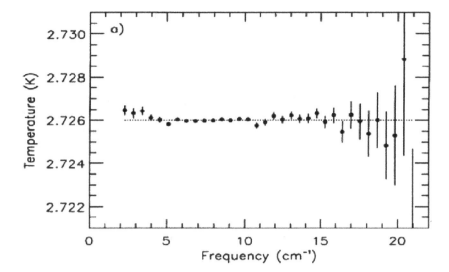

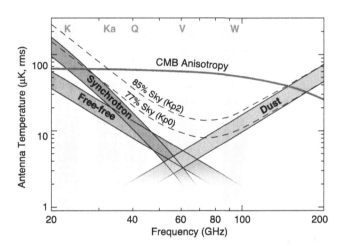

Fig. 4.4 Overview of the foreground spectra: the WMAP frequency bands were chosen to be in a spectral region where the CMB anisotropy (see Sect. 8) is dominant over the competing galactic and extragalactic foreground emissions (curves from http://lambda.gsfc.nasa.gov/product/foreground)

ing long-wavelength measurements of the spectrum. Moreover, as we shall see below, some of the spectral distortions we are looking for are more visible on the Rayleigh–Jeans side of the spectrum.

Finally, the FIRAS Team gave us, with the above black-body spectrum and its temperature, two upper limits on possible spectral distortions. But, now, let us discuss the various processes which could occur in the early universe and which could contribute to these different possible spectral distortions.

4.3 Thermalization

The CMB exhibits a perfect black-body shape to the precision of current measurements. The question arises: how did the black-body form and how was it maintained? In the early universe, a delicate interplay between radiation and matter took place. Moreover, any energy released into the primordial plasma could generate spectral distortions in the cosmic background radiation and, at present, signatures of these early epochs.

The main physical processes which were important for thermalizing the CMB spectrum in the early universe were Compton scattering:

$$e^- + \gamma \rightleftharpoons e^- + \gamma, \tag{4.13}$$

bremsstrahlung, also called free–free emission:

$$e^- + X \rightleftharpoons e^- + X + \gamma, \tag{4.14}$$

where X is an ion, and double-Compton scattering:

$$e^- + \gamma \rightleftharpoons e^- + \gamma + \gamma. \tag{4.15}$$

Their joined action changes the photon occupation number $\eta(\nu, t)$ with time [123]:

$$\frac{\partial \eta}{\partial t} = \frac{\partial \eta}{\partial t}\bigg|_C + \frac{\partial \eta}{\partial t}\bigg|_B + \frac{\partial \eta}{\partial t}\bigg|_{DC}. \tag{4.16}$$

Kompaneets [132] solved the general problem for the Compton scattering passing from the collisional integral equation to a very simple kinetic equation, which reads

$$\frac{\partial \eta}{\partial t}\bigg|_C = \frac{1}{t_C} \frac{1}{x_e^2} \frac{\partial}{\partial t}\left[x_e^4 \frac{\partial \eta}{\partial x_e} + \eta + \eta^2\right], \tag{4.17}$$

where $x_e = h_P \nu/(k_B T_e)$ is a dimensionless frequency; T_e is the electron temperature. The characteristic time t_C is given by

$$t_C \sim \frac{10^{28}}{\Omega_b h^2} \frac{2.725 \text{ K}}{T_0} \frac{T_r}{T_e}(1 + z)^{-4} \text{ s}, \tag{4.18}$$

where $T_r = T_0(1 + z)$ is the radiation temperature; the ratio T_e/T_r plays an important role only in cases where the electron temperature is not tightly coupled to the photon temperature.

Note that many authors have studied the problem for the three processes; in particular, Danese and De Zotti [123, 130, 131], Sunyaev and Zel'dovich [122], Burigana et al. [133, 134], Salati [41, 124], Hu and Silk [137], Danese and Burigana [135].

The bremsstrahlung term $\partial \eta/\partial t|_B$ is given by

$$\left.\frac{\partial \eta}{\partial t}\right|_B = \frac{1}{t_B} g_B(x_e) e^{-x_e} \frac{1}{x_e^3} \left[1 - \eta(e^{x_e} - 1)\right]; \qquad (4.19)$$

a quite similar expression holds for the double-Compton scattering term $\partial \eta/\partial t|_{DC}$:

$$\left.\frac{\partial \eta}{\partial t}\right|_{DC} = \frac{1}{t_{DC}} g_{DC}(x_e) e^{-x_e} \frac{1}{x_e^3} \left[1 - \eta(e^{x_e} - 1)\right], \qquad (4.20)$$

where $g_B(x_e)$ and $g_{DC}(x_e)$ are the Gaunt factors of bremsstrahlung and double Compton respectively, see for instance [136].

In terms of cosmological parameters the characteristic times during which the bremsstrahlung t_B and double Compton t_{DC} are able to change the occupation number are given by

$$t_B \sim \frac{10^{23}}{\Omega_b^2 h^4} \left(\frac{T_0}{2.725\,\mathrm{K}}\right)^{7/2} \left(\frac{T_e}{T_r}\right)^{7/2} (1 + z)^{-5/2}\,\mathrm{s}, \qquad (4.21)$$

$$t_{DC} \sim \frac{10^{39}}{\Omega_b h^2} \left(\frac{2.725\,\mathrm{K}}{T_0}\right)^2 \left(\frac{T_r}{T_e}\right)^2 (1 + z)^{-5}\,\mathrm{s}. \qquad (4.22)$$

Salati [124] has studied the typical time scales of these processes: t_C, t_B, t_D as a function of the redshift and has compared them to the age of the universe t_U in the case of an $\Omega = 1$ universe.

4.4 Spectral distortions of CMB radiation

Coupling between radiation and matter can lead to observable distortions of the CMB. If a significant amount of energy is injected to matter by some process, the coupling will transfer part of this energy into the CMB. From the study of Salati [124], one can show that three possibilities occur, depending on when the energy is released into the primordial plasma.

• $z \geq z_{Pl}$.

In this case the age of the universe t_U exceeded both t_C and t_{DC}. Double-Compton emission acted efficiently, and any spectral distortion was erased during an expansion time. No matter how large the energy release was, the microwave background is not affected. One can note that for $z \geq z_{Pl}$, the CMB radiation spectrum is not telling much about the

energetics of the universe. However, one can use big-bang nucleosynthesis to probe the total energy of the universe up to $z \sim 10^{10}$ (when the universe was 100 s old).

• $z_{BE} < z < z_{Pl}$.

The age of the universe t_U was still larger than the Comptonization time scale t_C, so that thermalization of the photon gas ensued. However, there is no production of additional photons because the double-Compton time scale t_{DC} exceeded t_U. Any energy release generated a μ-distortion, since the radiation spectrum relaxed towards a Bose–Einstein distribution faster than the universe expanded ($t_C < t_U$).

However, pure black-body radiation could not be achieved, since double-Compton emission was slower than expansion ($t_{DC} > t_U$). The photon bath was still in thermal equilibrium, but with an average photon energy larger than for a Planck spectrum, i.e. with a non-zero chemical potential μ. From the Bose–Einstein distribution, we have

$$\eta_\gamma^{BE} = \frac{1}{\exp(h_P \nu/k_B T_e - \mu/k_B T_e) - 1}. \qquad (4.23)$$

If $\Delta\rho_\gamma$ denotes the increase of the energy density due to an energy injection, one may express the chemical potential μ as a function of the relative increase of the photon energy by

$$\frac{\Delta\rho_\gamma}{\rho_\gamma} \sim 0.714\mu. \qquad (4.24)$$

• $z_{rec} \sim 10^3 < z < z_{BE}$.

Both t_{DC} and t_C exceeded the expansion time. Photons may still collide with electrons, and y-distortions may result if the Compton parameter y is small; $y = t_U/t_C$. If energy is released in the primordial plasma, the Rayleigh–Jeans region of the spectrum is slightly depleted by a factor $\exp(-2y)$, while a rapid rise in the temperature in the Wien region occurred. The magnitude of the distortion is related to the energy transfer $\Delta\rho_\gamma$ by

$$\frac{\Delta\rho_\gamma}{\rho_\gamma} \sim 4y. \qquad (4.25)$$

Therefore, one can say that the thermalization redshift is approximately given by $z_{Pl} \sim 10^7$. Any energy deposition at a later time, i.e. for $z \leq z_{Pl} \sim 10^7$, $T < 10$ keV or $t > t_{Pl} \sim 10^5$ s will leave a potentially observable distortion in the CMB spectrum:

• a μ-distortion for an early energy release: $z \sim 10^7$–10^5 with a possible drop in brightness temperature at centimeter wavelengths;

• a y-distortion for a late energy release: $z \sim 10^5$–10^3 with a decrease in temperature on the Rayleigh–Jeans side of the spectrum and a rise in temperature on the Wien side of the spectrum, i.e. a shift of photons from low to high frequencies. Therefore, precision measurements of the CMB spectrum probe the thermal history of the universe back to this epoch.

At present the definitive limits on both y and μ come from the FIRAS instrument aboard COBE. The 95% limits are [119]:

$$y \leq 1.5 \times 10^{-5}, \tag{4.26}$$

$$\mu \leq 9 \times 10^{-5}. \tag{4.27}$$

Therefore, the limit on energy release corresponding to a y-distortion for a late energy release ($z \sim 10^5 - 10^3$) is

$$\frac{\Delta\rho_\gamma}{\rho_\gamma} \leq 6 \times 10^{-5}, \tag{4.28}$$

and the limit on energy release corresponding to a μ-distortion for an early energy release ($z \sim 10^7 - 10^5$) is

$$\frac{\Delta\rho_\gamma}{\rho_\gamma} \leq 6.6 \times 10^{-5}. \tag{4.29}$$

Finally, from these limits on the energy release as a function of the redshift, one can induce limits on various processes which possibly took place in the early universe. We just mention:

- particle decays [124, 125];
- gravitational energy from large-scale structures [126, 127];
- primordial black-hole evaporation;
- superconducting cosmic strings and explosive formation [128, 129];
- variations of fundamental constants.

5 Cosmological recombination

In the hot big-bang picture, after the nucleosynthesis period a natural question concerns the possibility to form atoms by recombination of a primordial nucleus with free electrons, and a second question concerns the degree to which the recombination is inhibited by the presence of recombination radiation.

The reason why these questions are central were mentioned by Peebles [138] and Peebles and Dicke [139], based, at this epoch, on the *primeval-fireball* picture of Gamow [87] for the evolution of the universe.

The cosmological recombination process was not instantaneous, because the electrons, captured into different atomic energy levels, could not cascade instantaneously down to the ground state. Atoms reached the ground state either through the cosmological redshifting of the Lyman α line photons or by the $2s-1s$ two-photon process. Nevertheless the universe expanded and cooled faster than recombination could be completed, and a small fraction of free electrons and protons remained.

5.1 Recombination of hydrogen and deuterium

Peebles [138] was the first to present a theory in which the very complicated recombination process is reduced to simpler terms. The methodology of the calculations consists in taking into account a three-level atom with a ground state, a first excited state and a continuum, represented by a recombination coefficient. A single ordinary differential equation is derived to describe the ionization fraction. The assumptions are as follows.

- Hydrogen excited states are in equilibrium with the radiation.
- Stimulated de-excitation is negligible.
- Collisional processes are negligible.

The approximations are based on the rate of transfer of energy per unit volume between radiation and free electrons, developed by Kompaneets [132] and next by Weymann [140, 141] and finally coupled with the expansion of the universe.

One of the key points is to estimate the photorecombination rate R_{ci} directly to each level for multilevel H and D:

$$R_{ci} = 4\pi \int_{\nu_0}^{\infty} \frac{\sigma_{ic}(\nu)}{h_P \nu} I(\nu, T_r) \, d\nu. \tag{5.1}$$

Here i refers to the ith excited state and c refers to the continuum; ν_0 is the threshold frequency for ionization from the ith excited state. The radiation field $I(\nu, T_r)$, the Planck function, depends on the frequency ν and time t; see (4.7).

By using the principle of detailed balance in the case of local thermodynamical equilibrium (LTE), the radiative recombination rate can be calculated from the photoionization rate. Moreover, radiative recombination includes spontaneous and stimulated recombination.

Spontaneous recombination involves a free electron, but its calculation requires no knowledge of the local radiation field, because the energy of the photon is derived from the electron's kinetic energy [27]. The correction for stimulated recombination depends on T_m, because the recombination process is collisional. Thus the form of the total photoionization rate is

$$\sum_{i>1}^{N} n_i R_{ic} = \sum_{i>1}^{N} n_i 4\pi \int_{\nu_0}^{\infty} \frac{\sigma_{ic}(\nu)}{h_P \nu} I(\nu, T_r) \, d\nu, \tag{5.2}$$

and the total recombination rate is

$$\sum_{i>1}^{N} n_e n_c R_{ci} = n_e n_i \sum_{i>1}^{N} \left(\frac{n_i}{n_e n_c}\right)^{LTE}$$
$$\times 4\pi \int_{\nu_0}^{\infty} \frac{\sigma_{ic}(\nu)}{h_P \nu} \left(\frac{2h_P \nu^3}{c^2} + I(\nu, T_r)\right)$$
$$\times \exp\left(-\frac{h_P \nu}{k_B T_m}\right) d\nu. \tag{5.3}$$

5.2 Recombination of helium

The ^{4}He nucleus is the second most abundant nuclei in the primordial universe after the hydrogen nucleus. The conditions for the helium recombination are such that in both cases it occurs in accordance with the Saha equation [142, 143]. The recombination history of the early universe consists of three stages, but helium recombination occurs in two steps.

- at $z \sim 6\,000$ (or when the temperature became less than $16\,000$ K) the first electron recombines in order to form the singly ionized HeII from doubly ionized HeIII.
- Then at $z \sim 2\,700$ (or temperature close to $7\,300$ K), HeII recombines into a neutral state.
- from $z \sim 1\,500$ recombination of hydrogen took place.

Seager et al. [27] and then Dubrovich and Grachev [144] have developed an improved recombination calculation of H, He (with HeII and HeIII) that involves a line-by-line treatment of each atomic level. They found that HeI recombination is much slower than previously thought, and it is delayed until just before H recombines.

5.3 Recombination of lithium

The ionization potential $I_{Li} = 5.392$ eV of LiI is lower than that of hydrogen ($I_H = 13.6$ eV). Stancil et al. [149] suggested the existence of Li$^+$ ions after the period of the hydrogen recombination and showed that radiative recombination is again effective at the redshift $z \sim 450$. This last point is particularly important for lithium chemistry as we shall see.

6 Standard chemistry

The studies on the primordial chemistry (or postrecombination chemistry) have been the source of a tremendous increase of the literature. The first chemical network study including the primordial molecules (such as H_2, HD and LiH) and ions was carried out by Lepp and Shull [145], Latter and Black [146], Puy et al. [147], Galli and Palla [148], Stancil et al. [149], Lepp et al. [150], Pfenniger and Puy [151], and more recently Puy and Pfenniger [152].

After hydrogen recombination, the ongoing physical reactions are numerous, partly due to the presence of the cosmic microwave background radiation. Three classes of reactions are established.

- Collisional $\xi + \xi' \longleftrightarrow \xi_1 + \xi_2$ association, associative detachment, mutual neutralization, charge exchange, and reverse reactions.
- Electronic $\xi + e^- \longleftrightarrow \xi_1 + \xi_2$ recombination, radiative attachment, dissociative attachment.

- Photoprocesses $\xi + \gamma \longleftrightarrow \xi_1 + \xi_2$ dissociation, detachment, ionization, radiative association.

The chemical kinetics of a reactant ξ, which leads to the products ξ_1 and ξ_2, imposes the following evolution of the *mean numerical density* $\bar{n}_\xi$:

$$\left(\frac{d\bar{n}_\xi}{dt}\right)_{chem} = \sum_{\xi_1 \xi_2} k_{\xi_1 \xi_2} \bar{n}_{\xi_1} \bar{n}_{\xi_2} - \sum_{\xi'} k_{\xi \xi'} \bar{n}_\xi \bar{n}_{\xi'}, \qquad (6.1)$$

where $k_{\xi_1 \xi_2}$ is the reaction rate of the ξ-formation process from ξ_1 and ξ_2, and $k_{\xi \xi'}$ is the reaction rate of ξ-destruction by collision with the reactant ξ'. The typical rate of the collisional reactions, k_{coll} (in cm^3 s^{-1}), is calculated by averaging the cross sections, $\sigma(E)$, over a Maxwellian velocity distribution at temperature T_m:

$$k_{coll} = \sqrt{\frac{8}{\pi m_r (k_B T_m)^3}} \int_0^\infty \sigma(E) e^{-E/k_B T_m} E \, dE, \qquad (6.2)$$

where E is the collision energy, m_r the reduced mass of the collisional system (ξ, ξ'), and T_m is the temperature of the matter. The radiative rate coefficient k_{rad} (in s^{-1}), which depends on the radiative cross section σ_{rad}, and on the CMB radiation, is defined by

$$k_{rad} = \frac{8\pi}{c^2} \int_{\nu_{th}}^\infty \frac{\sigma_{rad}(\nu)\nu^2 \, d\nu}{\exp(h_P \nu / k_B T_r) - 1}, \qquad (6.3)$$

where ν_{th} is the threshold frequency above which the radiative process is possible, and T_r is the radiation temperature. As usual, c is the speed of light and h_P the Planck constant.

6.1 Helium chemistry

Helium chemistry plays an important role for the primordial chemistry because He is the first neutral atom which appeared in the universe. Once neutral He is significantly abundant, charged transfer with ions is possible, allowing for the formation of other neutral species. Thus HeH$^+$ is the first stable molecular bond formed in the radiative processes [153]:

$$He^{++} + e^- \rightarrow He^+ + \gamma, \qquad (6.4)$$

$$He^+ + e^- \rightarrow He + \gamma, \qquad (6.5)$$

$$He + H^+ \rightarrow HeH^+ + \gamma. \qquad (6.6)$$

The dominant destruction process at low redshifts is charge exchange via

$$HeH^+ + H \rightarrow He + H_2^+, \qquad (6.7)$$

which yields chemical equilibrium. However, knowledge of helium chemistry is still incomplete. For example, no information is yet available concerning HeD$^+$ [150].

6.2 Hydrogen chemistry

In the absence of solid surfaces, such as dust grains, it is possible to form neutral molecules through the radiative association between two neutral atoms. Molecular hydrogen cannot form easily by this radiative process, because H_2 does not have a permanent dipole moment.

Thus, any H_2 formed in the uniform background is dissociated by the radiation, until the density is too low to produce it. Nevertheless, once HeH^+ appears, it becomes an important source of H_2^+ through the exchange reaction with some neutral H, as we have mentioned in the preceding section. Charge transfer from H_2^+ becomes the most likely alternative to form H_2. Saslaw and Zipoy [154] and then Shchekinov and Entél [155] showed the importance of the charge-transfer reactions:

$$H_2^+ + H \rightarrow H_2 + H^+, \tag{6.8}$$

initiated by the radiation association

$$H + H^+ \rightarrow H_2^+ + \gamma. \tag{6.9}$$

They pointed out that the H_2^+ ion is converted to H_2 as soon as it is formed, and the H_2^+ concentration never becomes large.

Moreover, as the radiation temperature decreases, H_2 can be formed through H^- by radiative attachment. Peebles [138], in a scenario concerning the origin of globular star clusters, showed that these clusters may have originated as gravitationally bound gas clouds before the galaxies formed and suggested that some molecular hydrogen could be formed inside them, mainly by way of negative hydrogen by the reactions such as [155, 156])

$$H + e^- \rightarrow H^- + \gamma, \tag{6.10}$$

followed by the associative detachment

$$H^- + H \rightarrow H_2 + e^-. \tag{6.11}$$

The photodetachment of H^- and the photodissociation of H_2^+ by the background radiation field restrict the abundance of molecular hydrogen formed at the early stages, although the photodestruction of molecular hydrogen is negligible. The destruction is due to collisional dissociation. We notice that H_3^+ reactions, which are pivotal to the chemistry of dense interstellar clouds [157–159] do not play an important role in primordial chemistry.

6.3 Deuterium chemistry

HD forms in ways similar to H_2,

$$D^+ + H \rightarrow HD^+ + \gamma, \tag{6.12}$$

$$HD^+ + H \rightarrow H^+ + HD. \tag{6.13}$$

Deuterated-hydrogen molecules do have permanent dipole moments, which lead to the capacity to be formed by radiative association (forbidden between two hydrogen atoms):

$$H + D \rightarrow HD + \nu. \tag{6.14}$$

Nevertheless, HD formation is very significant when H_2 appears, the mechanism of dissociative collision

$$H^+ + D \rightarrow D^+ + H \tag{6.15}$$

$$D^+ + H_2 \rightarrow H^+ + HD. \tag{6.16}$$

Thus the formation of HD is carried out in two steps.

Nevertheless, when the abundance of H_2 is sufficient, the second way of formation is dominant (i.e. H^- way); see the work of Stancil et al. [149], who presented a complete review of the deuterium chemistry of the early universe.

6.4 Lithium chemistry

The lithium chemistry received large attention by the cosmochemistry community. Precise investigations of the chemical network were developed by [160] and [161]. Besides the classical radiative association, LiH can be formed by different exchange reactions. The lithium chemistry is initiated by the recombination of lithium, which occurred near the redshift $z \sim 450$. The formation of the molecular ion LiH^+, formed by radiative association processes,

$$Li^+ + H \rightarrow LiH^+ + \gamma, \tag{6.17}$$

$$H^+ + Li \rightarrow LiH^+ + \gamma, \tag{6.18}$$

opens the way of the formation of the LiH molecules through the exchange reactions:

$$LiH^+ + H \rightarrow LiH + H^+, \tag{6.19}$$

which are more rapid than the formation by radiative association of H and Li atoms. Despite the radiative association from electronically excited lithium atoms, tentatively proposed by [162–164], the LiH abundance remains low.

6.5 Equations of evolution

All the rates of recombination and chemical reactions depend on the electron density n_e, the matter and radiation temperatures and the baryon density. Thus, it is necessary to take into account the dynamical and thermal equations of evolution.

6.5.1 Evolution of matter and radiation temperature

The matter temperature is affected by the Compton cooling (which is the major source of energy transfer between electrons and photons), adiabatic cooling (due to the expansion of the universe).

The first law of thermodynamics imposes

$$\frac{d}{dt}[\varepsilon_{tot}V] = -p_{tot}\frac{dV}{dt} + \Gamma_{com}V, \qquad (6.20)$$

p_{tot} is the total pressure, which depends on p_r (radiative pressure) and p_m (baryonic gas pressure):

$$p_{tot} = p_r + p_m = \frac{a_r T_r^4}{3}(1 + f_\nu) + \bar{n}_b k_B T_m, \qquad (6.21)$$

where f_ν is the neutrino contribution to the radiation density for three massless, non-degenerate, neutrino flavors given by

$$f_\nu = 3\frac{7}{8}\left(\frac{4}{11}\right)^{4/3}. \qquad (6.22)$$

Γ_{com} is the energy transfer from the radiation to matter via Compton scattering of CMB radiation on free electrons [26, 140], given by

$$\Gamma_{com} = 4\bar{n}_e \sigma_T a_r k_B T_r^4 \frac{T_r - T_m}{m_e c}, \qquad (6.23)$$

and σ_T defines the Thomson cross section, m_e the electron mass and, finally, $\bar{n}_e$ the mean numerical density of the electron.

Note that in the external heating and cooling contribution, the sum of enthalpy reaction and energy transfer (via excitation and de-excitation of the molecular transition) are negligible [147, 148, 150] (but this is not the case in the collapse of molecular protoclouds).

Thus, in a comoving frame the evolution of the baryon temperature T_m is given by the expression

$$\frac{dT_m}{dt} = -2H(t)T_m + \frac{8}{3}\frac{\sigma_T a_r}{m_e c}T_r^4(T_r - T_m)x_e, \qquad (6.24)$$

where $H(t)$ is the Hubble parameter, $x_e = \bar{n}_e/\bar{n}_b$ characterizes the ionization fraction (i.e., the electronic abundance).

This equation is coupled with the radiation temperature evolution and the time–redshift relation:

$$\frac{dT_r}{dt} = -H(t)T_r. \qquad (6.25)$$

6.5.2 Evolution of the baryon density

The numerical mean baryon density, $\bar{n}_b$, is expressed by

$$\bar{n}_b = \sum_\xi \bar{n}_\xi = \frac{\sum_\xi N_\xi}{V}, \qquad (6.26)$$

where N_ξ is the number of species ξ, and V is a comoving volume proportional to the scale factor cubed, $V \propto a^3$. Thus the mean numerical density evolution is given by

$$\frac{d\bar{n}_b}{dt} = -3H\bar{n}_b + \sum_\xi \left(\frac{d\bar{n}_\xi}{dt}\right)_{chem}. \qquad (6.27)$$

The last term of the second member of this equation represents the variation of the number of particles due to chemistry.

Moreover, for each chemical species ξ we have

$$\frac{d\bar{n}_\xi}{dt} = -3H(t)\bar{n}_\xi + \left(\frac{d\bar{n}_\xi}{dt}\right)_{chem}, \qquad (6.28)$$

where $\bar{n}_\xi$ is the numerical density of ξ. The last term of the second member of this equation expresses the contribution of the chemical kinetics, see (6.1), and H is the Hubble parameter given by (1.2).

6.6 Results of thermochemistry

The set of ordinary differential equations depends on the temperatures (radiation or matter) and the matter density.

We have solved the full equations for the time evolution of the chemical abundance, temperature and baryonic density with the numerical code developed by Puy and Pfenniger [152].

Here we use the chemical network and reaction rates described by the seminal papers of [148] and [165], except for the hydrogen and deuterium recombinations which are calculated from the reaction rates given by [166] from numerical approximations.

6.6.1 Cosmological recombination

Various approaches exist to the calculation of the time evolution of the electronic fraction during the cosmological recombination; see the exhaustive work of Schleicher et al. [167].

In our case, in order to illustrate the mechanism of cosmological evolution, the value of the rate of radiative transition from the continuum to the excited states is the simple power-law fit given by [166].

The results are shown in Fig. 6.1, separately for each atomic element (i.e. H, He, D and Li). We introduce the redshift of the recombination z_{rec} by the epoch when the abundance of a species equals the one of its corresponding ion. Thus, we obtain the successive redshifts of HeII, HeI, DI and HI recombination (see Fig. 6.1):

- $z_{rec,He^+} \approx 6\,101$,
- $z_{rec,He} \approx 2\,604$,
- $z_{rec,D} = z_{rec,H} \approx 1\,425$.

Fig. 6.1 Successive
cosmological recombination
versus redshift. The helium and
hydrogen recombination are
plotted on the *right panel*, while
the deuterium and lithium
recombination are indicated on
the *left panel*. The ordinates
represent the relative
abundances, calculations from
the numerical code developed
by Puy and Pfenniger [152]

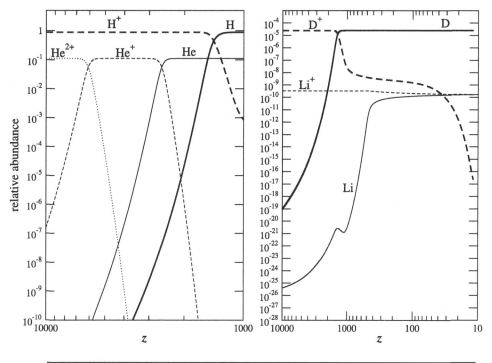

Fig. 6.2 Evolution of the
chemical abundances of helium
components. The redshift of the
successive helium
recombination are indicated.
The ordinates represent the
relative abundances,
calculations from the numerical
code developed by Puy and
Pfenniger [152]

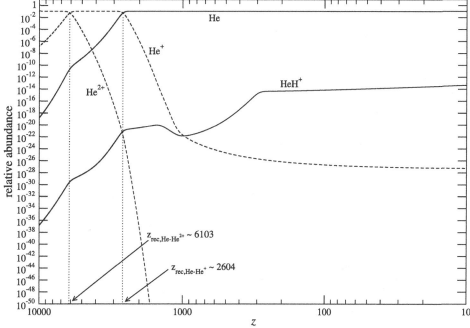

Let us notice that the ionization energy of neutral lithium amounts to $I_{Li} \sim 5.39$ eV, at the lowest ground-state configuration. Despite the more abundant electrons after hydrogen recombination (a few times 10^{-4}, while Li^+ reaches a few times 10^{-10}), Li might be the last element that we consider to recombine. However, the charge transfer (Li–H^+) and (Li^+–H) remains active. For these reasons Li does not recombine totally, and neutral and ionized Li tend to match their abundances.

We find that the residual ionization fraction x_e is close to 5.78×10^{-5}. The reduced number of free electrons and

protons are going to have a direct effect on the chemical evolution of the gas.

6.6.2 Helium chemistry

The main molecular species containing helium is HeH^+, formed, as we have seen, by the radiative association processes; see Fig. 6.2. As Galli and Palla [148] mentioned, the HeH^+ ions are removed by the CMB photons at $z < 250$ also by collisions with H atoms. This process explains the abrupt change in the slope of the curve of HeH$^+$, as shown in Fig. 6.2.

At the redshift $z_f = 10$ we find the *frozen* abundances:

- [He] = 0.111,
- [He$^+$] = 6.41×10^{-28},
- [He^{++}] = 6.86×10^{-96},
- [HeH$^+$] = 4.6×10^{-14}.

6.6.3 Hydrogen chemistry

Figure 6.3 illustrates the evolution of the hydrogen species. We see that the formation of H$_2^+$ is controlled by radiative association and photodissociation at $z > 100$. At high redshift, the formation of this ion is regulated by the destruction of HeH$^+$.

The steady drop of H$^-$ is determined by the mutual neutralization with H$^+$ at $z < 100$.

Thus, the two major routes of H$_2$ formation are relative to the two jumps at the redshifts $z \sim 500$ (via the H$_2^+$ channel) and $z \sim 150$ (via H$^-$ channel).

At the redshift $z_f = 10$ we find the hydrogen abundances

- [H] = 0.889,
- [H$^+$] = 5.78×10^{-5},
- [H$^-$] = 7.34×10^{-14},
- [H$_2^+$] = 6.63×10^{-14},
- [H$_3^+$] = 3.55×10^{-17},
- [H$_2$] = 1.13×10^{-6}.

6.6.4 Deuterium chemistry

Figure 6.4 shows the evolution of deuterium chemistry. We remark that the only molecule formed in significant amounts is HD, whose evolution with redshift follows closely that of H$_2$. A complete description of the abundance evolution is found in the paper of Galli and Palla [148].

The final deuterium abundances are, at the redshift $z_f = 10$,

- [D] = 2.436×10^{-5},
- [D$^+$] = 2.1×10^{-19},
- [HD$^+$] = 5.31×10^{-20},
- [H$_2$D$^+$] = 2.33×10^{-19},
- [HD] = 3.67×10^{-10}.

6.6.5 Lithium chemistry

The abundances of lithium chemistry are very low as we see in Fig. 6.5. The more abundant complexes are LiH and LiH$^+$, whose the formation is sensibly controlled by the radiative association of Li$^+$ and H (see [160] and [148] for an exhaustive description).

The final deuterium abundances are, at the redshift $z_f = 10$,

- [Li] = 1.668×10^{-10},
- [Li$^+$] = 1.675×10^{-10},
- [Li$^-$] = 1.25×10^{-21},
- [LiH$^+$] = 1.26×10^{-18},
- [LiH] = 2.53×10^{-20}.

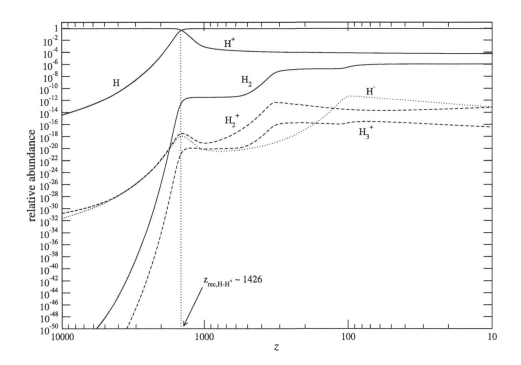

Fig. 6.3 Evolution of the chemical abundances of hydrogen components. The redshift of the hydrogen recombination is indicated. The ordinates represent the relative abundances, calculations from the numerical code developed by Puy and Pfenniger [152]

Fig. 6.4 Evolution of the chemical abundances of the deuterium components. The redshift of the deuterium recombination is indicated. The ordinates represent the relative abundances, calculations from the numerical code developed by Puy and Pfenniger [152]

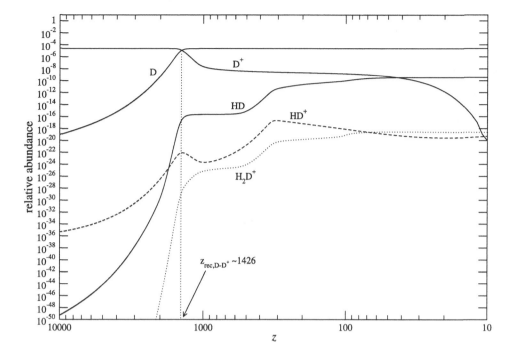

Fig. 6.5 Evolution of the chemical abundances of lithium components. The ordinates represent the relative abundances, calculations from the numerical code developed by Puy and Pfenniger [152]

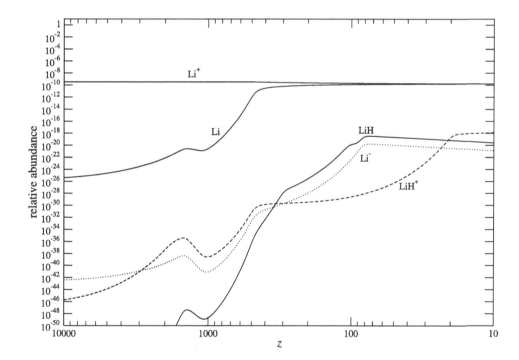

6.6.6 Thermal decoupling

Due to the decreasing of the ionized fraction, Thomson scattering becomes more and more inefficient during the evolution of the universe. Actually the temperature of matter (i.e. without radiation coupling) evolves as $(1+z)^2$ when the radiation temperature evolves as $1+z$. We can define an ef-

fective redshift of decoupling $z_{dec,eff}$ by *last* epoch when the radiation and matter temperatures are equal:

$$T_r(z_{dec,eff}) = T_m(z_{dec,eff}). \tag{6.29}$$

Moreover, we define the decoupling redshift $z_{dec,1\%}$, for which the matter temperature is equal to 1% of the radiation temperature.

Fig. 6.6 Thermal decoupling between radiation and matter in the standard big-bang chemistry model. T_r (*dashed line*) represents the temperature of radiation, T_m (*bold line*) is the temperature of matter. T_{ms} (*thin line*) is the *standard* matter evolution, where $T_{m0} = T_{r0}/(1 + z_{dec,eff})$. $z_{dec,\xi}$ is respectively the redshift of recombination for He$^+$, He, H and D. $z_{dec,1\%}$ is the decoupling redshift for which the matter temperature is equal to 1% of the radiation temperature; figures from Puy and Pfenniger [152]

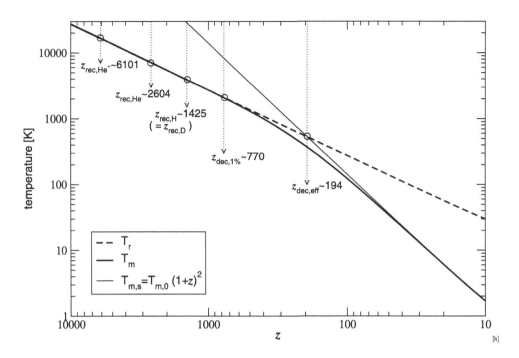

Figure 6.6 represents the evolution of matter and radiation. We see that the two epochs, $z_{dec,eff}$ and $z_{dec,1\%}$, are

$$z_{dec,eff} \sim 194 \quad \text{when } z_{dec,1\%} \sim 770, \tag{6.30}$$

which differ from the common approximation $z_{dec} \sim 1\,500$. Generally there is a strong confusion between the redshift of hydrogen recombination—we have seen $z_{rec,H} \sim 1\,425$—and this redshift of thermal decoupling.

Let us notice that at the redshift $z_{dec,1\%} \sim 770$, matter is again strongly thermally coupled with the radiation. This is, in part, the reason that the formation of the first objects will be later.

7 The CMB dipole anisotropy

As already said in Sect. 1, the photons in the universe decoupled from matter at $T \sim 3\,000$ K or $z \sim 1\,000$. These photons have been propagating freely in space-time since then and can be detected today.

In an ideal Friedmann universe, a comoving observer will see these photons as a black-body spectrum at the temperature T_0. The deviations in the metric from that of the Friedmann universe, the motion of the observer with respect to the comoving frame and any astrophysical process occurring along the trajectory of the photon can lead to effects in this radiation. There are two possible kinds of effects.

(1) From a particular direction in the sky, the spectrum may not be strictly Planckian with a single temperature T_0.

(2) The spectrum, in any direction, may be Planckian, but the temperature may be different in different directions:

$T_0 = T_0(\theta, \phi)$, where θ and ϕ are the angular coordinates on the sky.

In Sect. 4, we have seen that there is no distortion of the first kind in the CMB. In particular, FIRAS aboard COBE gave only upper limits on possible μ- and y-distortions which could establish strong constraints on many cosmological models.

On the contrary, the CMB radiation shows deviations of the second kind at different angular scales generally called anisotropies. These CMB anisotropies are the subject of Sects. 7, 8 and 9.

The most dominant anisotropy in the CMB radiation, which could be interpreted as being due to the motion of our galaxy with respect to the cosmic rest frame, would exhibit a so-called dipole pattern, hotter in our direction of motion and colder in the opposite. Partly following previous discussions [19, 171], we shall present this dipole anisotropy: its prediction, its discovery and measurements and its physical origin.

7.1 The prediction of the dipole anisotropy

7.1.1 The historical problem

The first step towards the study of the dipole anisotropy was done by Hasenöhrl [168, 169] and by von Mosengheil [170]. They computed the Doppler signal seen by an observer moving at a constant speed v inside a cavity at a given temperature T on the basis of Galilei and Lorentz transformations, respectively. The radiation emitted by the cavity is that of a black body. The problem is: how it will appear to the moving observer? For von Mosengheil—who was a student of

Max Planck—the proposal of his thesis combined two important discoveries: Planck's formula for a black body and the Lorentz transformations recently introduced by Einstein in his theory of special relativity. After a cumbersome computation, he found the correct result [170]:

$$T_{obs} = T_{wall} \sqrt{1 - \frac{v^2}{c^2}} [1 - \mathbf{v}.\mathbf{n}/c]^{-1}, \tag{7.1}$$

where the vector $\mathbf{n}$ gives the direction of observation. In von Mosengheil's time, there was no chance of verifying the solution through an experiment. This result had been forgotten until 1965, when the CMB was discovered.

7.1.2 The present predictions

In the framework of big-bang theory, we are living inside a *cavity* limited by the so-called *last scattering surface* (see Sect. 1), a layer of ionized gas, opaque to electromagnetic radiation.

The only difference with respect to the situation discussed by von Mosengheil is that the large scattering surface is expanding and receding from us, so that the photons of the last scattering surface suffer a substantial reddening, thereby appearing as a black body at about 2.7 K.

One notes that the recession of the large scattering surface is due to the expansion of space and it is not a pure velocity of recession: the proper velocity of the observer adds algebraically to the recession velocity and not through the Lorentz formula of special relativity. Since the earth is moving around the sun, the sun is moving inside the galaxy and the galaxy is expected to move with respect to other galaxies, an earth's observer should detect the effect noted by von Mosengheil, thereby measuring his motion with respect to the last scattering surface. In a first approximation in $\beta = v/c$, the above equation predicts a *dipole pattern* on the sky, i.e.

$$T_{obs} \sim T_{wall}(1 + \beta \cos \Theta) \quad \text{with } \beta = v/c, \tag{7.2}$$

where Θ is the angle between the direction of observation and that of the motion.

Later on, at the time of CMB observations, this pattern became known as the *dipole anisotropy*.

In early 1967, with only scarce and marginal data available on the isotropy of CMB—in particular, Partridge and Wilkinson [172] have established an upper limit of about 3 mK to any large-scale component of the CMB anisotropy—Sciama [173] noted: *It should be possible to measure the "absolute" motion of the solar system, that is, its motion relative to the universe as a whole.*

As a consequence, Sciama attempted to derive the motion of the sun versus distant galaxies and added the important note: that *in principle all the galaxies concerned (z < 0.5)*

could be moving as a whole with respect to the matter with larger redshifts.

Also in 1967, Rees and Sciama [174], and next Steward and Sciama [175], gave different predictions in approximate agreement with the *possible* detection suggested by Partridge and Wilkinson [172]. Next, all these predictions turned out to be in disagreement with observations, but they confirmed the validity of the cautious comment of Sciama about the possibility of a *bulk* motion. All these publications stimulated the interest of experimentalists toward direct observations of the CMB dipole. Several authors attempted to estimate the sky pattern due to this anisotropy.

In 1968, Heer and Kohl [176] and—in the same volume of Physical Review—Peebles and Wilkinson [177] published the correct result. In particular, they used the following expression:

$$T = T_0(1 + \beta \cos \Theta), \tag{7.3}$$

where T_0 is the mean temperature of the CMB over the sky, $\beta = v/c$ is the velocity of the observer with respect to the LSS at $z \sim 1\,000$, and Θ is the angle between the direction of observation and that of motion.

Finally, note that the derivation of the dipole formula can be done by briefly considering the distribution function $f(\mathbf{p})$ of particles of momentum $\mathbf{p}$ in the observer's frame (see [178] and all the good textbooks such as [179]). These quantities, $f(\mathbf{p})$ and $\mathbf{p}$, obey the following relation:

$$f(\mathbf{p}) = f'(\mathbf{p}'), \tag{7.4}$$

where $f'(\mathbf{p}')$ is the distribution function in the LSS frame in which the photon background is isotropic and where $\mathbf{p}$ and $\mathbf{p}'$ are related through the Lorentz transformation:

$$\mathbf{p} = \sqrt{1 - \frac{v^2}{c^2}} [1 - \mathbf{v}.\mathbf{n}/c]^{-1} \mathbf{p}'. \tag{7.5}$$

Applying these equations to a black body, one finds

$$T(\mathbf{n}) = \sqrt{1 - \frac{v^2}{c^2}} [1 - \mathbf{v}.\mathbf{n}/c]^{-1} T'. \tag{7.6}$$

Introducing Θ as the angle between the direction of motion and that of observation:

$$T(\Theta) = \sqrt{1 - \beta^2} [1 - \beta \cos \Theta]^{-1} T'. \tag{7.7}$$

This is the formula derived by von Mosengheil [170], Peebles and Wilkinson [177], Heer and Kohl [176].

7.2 The observation of the dipole anisotropy

First of all, we have to recall the observations of Bracewell and Conklin [180] spanning from 1967 to 1969 [181, 182].

They were carried out from the ground with a radiometer, working at 8 GHz and pointing at a constant declination with a peak at 13 h RA (right ascension). Galactic contamination was rather severe. These observations are described in the PhD thesis work of Conklin (performed at Stanford University, 1969). Later, during a IAU Symposium, Conklin reported more data, claimed dipole detection, estimating an amplitude of

$$\Delta T \sim 2.28 \pm 0.92 \text{ mK}, \tag{7.8}$$

with the peak at 10 h 58′ RA [183]. The declination of the dipole peak was first measured by Henry [184] for his PhD thesis work (Princeton University, 1971). Henry and Wilkinson used a balloon-borne radiometer at 10–15 GHz. They found

$$\Delta T = 3.2 \pm 0.8 \text{ mK}, \tag{7.9}$$

directed toward a RA of 10.5 ± 4 h and a declination D of $[-30 \pm 25]°$. These preliminary observations—Conklin in 1969 and in 1972, Henry in 1971—have indicated the existence of a dipole anisotropy; but both the direction and the amplitude were largely uncertain. The next generation of experiments started in 1976 with the balloon experiments of the Princeton group [185], the U-2 experiments of the Berkeley group [186] and the balloon experiments of the Florence group [187].

These three groups precisely measured the amplitude and direction of the dipole anisotropy and extended the analysis to millimeter wavelengths, thereby confirming the Planckian character of the observed signal. A first review of the subject is in [93], where the previous data are compared with the first results of COBE; see also [19, 171].

The main CMB dipole measurements—which are more and more accurate both in amplitude and in direction—are collected in Table 7.1.

Note that in this table the results from Henry and from Corey and Wilkinson constituted respectively the discovery and a first confirmation of the dipole, while the other

ones provided improved measurements of the dipole. Remark also that the results from the Melchiorri group were the first ones, before DMR/COBE, in the millimetric part of the CMB.

7.3 The physical origin of the dipole anisotropy

The large value—about two orders of magnitude higher than other anisotropies—of the dipole anisotropy clearly points in favor of a kinetic origin. The kinetic explanation requires a peculiar velocity of the observer, which can be only explained as the result of gravitational acceleration integrated in time.

Therefore, the dipole anisotropy results from the sum of many components of velocity due to gravitational attraction of various mass concentrations.

In particular, from the value of the dipole measured by WMAP [191], we have $\Delta T = 3.346 \pm 0.017$ mK, and the implied velocity for the solar system barycenter is $v = 368 \pm 2$ km/s, towards $(l, b) = (263.85° \pm 0.1°, 48.25° \pm 0.04°)$. Such a velocity of the solar system implies a velocity—relative to the CMB—for the galaxy and the Local Group of galaxies of $v_{LG} = 627 \pm 22$ km/s towards $(l, b) = (276° \pm 3°, 30° \pm 3°)$. For a discussion of the size of the region within which the motion of the Local Group is shared by other galaxies, see for instance [192]; Branchini, Plionis and Sciama [193] have observed that the gravitational acceleration of many different samples of extragalactic objects is relatively well aligned with the direction of the CMB dipole.

The motion is also confirmed by measurements of the velocity field of local galaxies [194]. Anyway, understanding the peculiar velocity field in the universe is an important addition to the study of the anisotropies in the CMB.

7.4 Conclusions

An examination of the CMB map—see for instance Fig. 7.1—reveals a very gradual shift in temperature, by

Table 7.1 The main CMB dipole measurements (dipole amplitude, galactic longitude l, galactic longitude b, observational frequency)

Reference	Amplitude $\pm 1\sigma$ (in mK)	$l \pm 1\sigma$	$b \pm 1\sigma$	Frequency (in GHz)
Henry [184]	3.3 ± 0.7	270 ± 30	24 ± 25	10
Corey and Wilkinson [185]	2.4 ± 0.6	306 ± 28	38 ± 20	19
Smoot et al. [186]	3.5 ± 0.6	248 ± 15	56 ± 10	53
Fabbri et al. [187]	2.9 ± 0.95	256.7 ± 13.8	57.4 ± 7.7	300
Boughn et al. [188]	3.78 ± 0.30	275.4 ± 3.9	46.8 ± 4.5	46
Lubin et al. [189]	3.44 ± 0.17	264.3 ± 1.9	49.2 ± 1.3	90
COBE/FIRAS [119]	3.372 ± 0.005	264.14 ± 0.17	48.26 ± 0.16	300
COBE/DMR [190]	3.358 ± 0.023	264.31 ± 0.17	48.05 ± 0.10	53
WMAP [191]	3.346 ± 0.017	263.85 ± 0.1	48.25 ± 0.04	

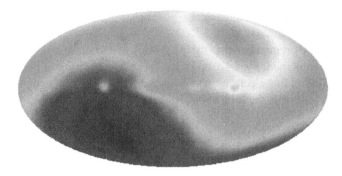

Fig. 7.1 Map of the dipole based on four years of DMR/COBE observations: the *red part* of the sky is hotter by $(v/c)T_0$, while the *blue part* of the sky is colder by $(v/c)T_0$. This all-sky image in galactic coordinates is plotted using the equal-area Molweide projection; see http://lambda.gsfc.nasa.gov/product/cobe

about 1.27 part in 1 000, from one side of the sky to the opposite side. One direction is a little hotter than the average, and the opposite direction a little cooler. The origin of this dipole anisotropy seems to be kinetic, thereby confirming the initial predictions of Sciama: maps of the CMB enable us to measure our velocity through the universe quite precisely—both how fast we are going and in what direction—to an accuracy of a few percent.

8 Cosmic microwave background anisotropies (CMBA)

The detection of the CMB radiation [7] and the confirmation of the thermal nature of the CMB radiation by the COBE/FIRAS instrument [195]) can be seen as the first revolution of the 2.7 K radiation.

The temperature anisotropy of the CMB, first detected by the COBE/DMR instrument [12]—spatial variations of order $\Delta T/T \sim 10^{-5}$ across $10°$–$90°$ on the sky—can be seen as the *second revolution* of the 2.7 K radiation.

These 10^{-5} variations in temperature represent the direct imprint of initial gravitational potential perturbations on the CMB photons, called the Sachs–Wolf effect [196].

One must note that until 1990 CMB anisotropies have been supposed to be marginally detected, in a few limited regions, by the Groups of Melchiorri [197], Lubin [199] and Davies [198].

The very low amplitude of the observed signals, usually at 1–2 standard deviations above the noise, and the lack of confirmations have led cosmologists to consider these observations as *upper limits*.

However, these upper limits, especially those at small angular scales by Wilkinson Group [200] and those around 3 and 6 degrees by Davies and Melchiorri have led cosmologists to exclude pure baryonic universes, thereby opening the door to the possibility of contributions of non-baryonic dark matter.

Then, just after the COBE/DMR detection, from 1992 to 1998, several experiments detected a rise and fall in the level of anisotropy from degree scales to arcminute scales.

From a theoretical point of view, since the seventies, cosmologists like Peebles and Yu [201], Doroshkevich et al. [202] and Bond and Efstathiou [203] have shown that most of the structure in the temperature anisotropy, on scales close on ~1° scales, must be associated with acoustic oscillations of the photon–baryon plasma at the time of recombination or just before.

The detailed theory predicting these CMBA—these acoustic *peaks* in the CMB temperature—has been elaborated by Seljack [204], Hu and Sugiyama [221], Hu and White [205] and Hu et al. [206], in particular.

The location of the first peak of the anisotropy was determined by the BOOMERANG [207], MAXIMA [208] and ARCHEOPS [31] experiments. The measurement of this first peak has been confirmed by many other groups. More precise measurements of the first five peaks are currently done by ground-based experiments and by the satellite WMAP [66, 217].

General reviews on the CMB anisotropies have been written, in the last decade, by many authors. Among them, let us recall Bond [209], Kamionkowski and Kosowski [210], Hu and Dodelson [211], Hu [213] and, more recently, again Hu [214].

Present and future CMB measurements are essentially concerned with small-scale anisotropies on scales of close to 1° and less. The recent data, WMAP in particular, provide undeniable evidence for the interpretation of these small-scale CMBA in terms of acoustic oscillations; therefore we adopt the notation, the set of equations governing the small-scale CMB anisotropies and, of course, the interpretation given by Hu et al.

In this section, we define the CMBA observables; then we recall the role of Thomson scattering in the primordial plasma. Therefore, we recall the basics of fluid dynamics, and finally, we present acoustic oscillations in the primordial plasma.

8.1 CMB observables

The primary aim of the CMB satellite experiments such as COBE, WMAP and PLANCK is to map the temperature $T(\check{\mathbf{n}})$ of the CMB and its polarization, described by the Stokes parameters $Q(\check{\mathbf{n}})$ and $U(\check{n})$, as functions of position $\check{\mathbf{n}}$ on the sky. Many temperature–polarization angular correlation functions, or equivalently, power spectra, can be extracted from such maps. Then these quantities can be compared with detailed predictions from cosmological models.

A harmonic description is well adapted in order to analyze the anisotropy of the distribution of photons in the sky.

Thus the temperature field $\Theta(\check{\mathbf{n}})$ can be expressed in spherical harmonics:

$$\Theta(\check{\mathbf{n}}) = \frac{\Delta T}{T} = \frac{T(\check{\mathbf{n}}) - T_0}{T_0} = \sum_{lm} \Theta_{lm} Y_{lm}(\check{\mathbf{n}}), \qquad (8.1)$$

$\check{\mathbf{n}}$ is the normalized direction of the line of sight, Y_{lm} is the ordinary harmonic and $T_0 \sim 2.728$ K is the present CMB temperature [200]. If we have a Gaussian random temperature fluctuations, the fluctuations are fully characterized by their power spectrum (where the * symbol characterizes the conjugate term):

$$\langle \Theta_{lm}^* \Theta_{l'm'} \rangle = \delta_{ll'} \delta_{mm'} C_l^{TT}, \qquad (8.2)$$

with

$$C_l^{TT} = \sum_{m=-l}^{m=l} \Theta_{lm} \Theta_{lm}{}^\star. \qquad (8.3)$$

The temperature correlation power spectrum, $\Delta_T |_{TT}$, is usually displayed thus:

$$\Delta_T^{TT} \equiv T \sqrt{\frac{l(l+1)C_l^{TT}}{2\pi}} \quad (\text{in } \mu K), \qquad (8.4)$$

the power per logarithmic interval in wave number for $l \gg 1$. This quantity gives the contribution to the total variance in temperature from the mean-square fluctuation at an angular scale: $\theta \sim 180°/l$. Figure 8.1, which shows the temperature power spectrum from WMAP in five-year data, provides *incontrovertible* evidence for the interpretation of these anisotropies in terms of *acoustic peaks* [217].

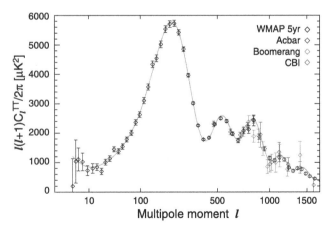

Fig. 8.1 The WMAP five-year TT power spectrum along with recent results from the ACBAR (due to Reichardt [215], *purple*), Boomerang (Jones [224], *green*), and CBI (Readhead [225], *red*) experiments. The red curve is the best-fit ΛCDM model to the WMAP data, (curves from the CAPMAP experiment; see http://lambda.gsfc.nasa.gov/product/map)

8.2 Thomson scattering

Let us remember that the properties of the plasma before recombination are governed by Thomson scattering of photons on free electrons, which has a differential cross section, where ϖ is the solid angle, such as

$$\frac{d\sigma}{d\varpi} = \frac{3}{8\pi} \left[\check{\mathbf{E}}' . \check{\mathbf{E}} \right]^2 \sigma_T, \qquad (8.5)$$

with the Thomson cross section

$$\sigma_T \sim 6.65 \times 10^{-25} \text{ cm}^2. \qquad (8.6)$$

$\check{\mathbf{E}}$ and $\check{\mathbf{E}}'$ denote the incoming and outcoming directions of the electric field.

The optical depth τ (related to the comoving mean free path of photons, l_λ), is given by

$$d\tau = n_e \sigma_T c \, dt. \qquad (8.7)$$

This equation is coupled with the evolution equation of chemistry, because it is strongly dependent on the electron densities (see Sect. 5).

At $z = 1\,000$, where the ionized fraction $x_e \sim 1$, the comoving mean free path of photons is $l_{\lambda 1000} \sim 2.5$ Mpc. On scales $l_\lambda \gg l_{\lambda 1000}$, photons are tightly coupled to the electrons by Thomson scattering, which in turn are tightly coupled to the baryons by Coulomb interactions. The dynamics that results involves for the primordial plasma description by a single photon–baryon fluid (called the *tight-coupled fluid* or the *photon-dominated fluid*) instead of two fluids for the general case.

Now, we shall consider inhomogeneities in the primordial plasma. These inhomogeneities, which will eventually become the large-scale structures we see in the universe today, grew from tiny seeds at the last scattering surface. To handle this situation, we need to use relativistic perturbation theory in an expanding universe. Well inside the horizon, we can use ordinary Newtonian gravity, but on larger scales the details of general relativity come into play. Details of relativistic perturbation theory lie beyond the scope of this paper but are discussed in [218–220].

Let us only remark that there exists a disparity between the smooth photon distribution and the clumpy matter distribution which is due to radiation pressure: even if inhomogeneities in the matter in the universe and anisotropies in the CMB originated from the same sources, they appear very different today. Here, as said above, we adopt the point of view of Hu et al. and treat the primordial plasma in a *fluid approximation*. We shall see that the photon pressure sets up sound waves or acoustic oscillations in the fluid. These sound waves are frozen into the CMB at recombination. Regions that have reached their maximal compression by recombination become hot spots on the sky; those that reach maximum rarefaction become cold spots. In order to understand the angular pattern of temperature fluctuations seen by the observer today, one must not only understand the spatial temperature pattern at recombination, but more generally, the evolution of perturbations in the photon–baryon fluid. In the next subsection, we shall recall some sound physics (i) in a perfect fluid, (ii) when gravity effects are added, (iii) when baryon effects are also taken into account, (iv) and finally, when the most dissipative baryon–photon fluid is considered.

8.3 Acoustic oscillations in the photon–baryon fluid

As suggested above, we consider, for the primordial plasma, a description by a photon–baryon fluid. Perturbations, in this description, satisfy an equation of continuity and an Euler equation—for each component—which can be written in Fourier space.

The spatial power spectrum at recombination can be decomposed into its Fourier modes:

$$\Theta(x) = \frac{1}{(2\pi)^3} \int e^{ikx} \Theta(k) \, d^3k, \tag{8.8}$$

where k is the wave number. Since perturbations are very small, we can do the linear approximation for the perturbations. We introduce the following variables: the photon "temperature" $\Theta = \Delta T/T$, the photon-fluid velocity v_γ for the photon fluid, the baryonic-density perturbation δ_b, the bulk velocity v_b for the baryon fluid. We consider that in the primordial plasma this photon-fluid velocity, in particular, is approximately parallel to the wave vector $\mathbf{k}$. In the following subsection, in the framework of the very particular case of a *perfect-fluid approximation* for the primordial plasma, we can show the main features—peaks and troughs—of the anisotropy spectrum.

8.3.1 Acoustic oscillations in the photon-dominated fluid

Therefore, we consider, first, the *tight-coupled fluid* or the *photon-dominated fluid* as a perfect gas without gravity and dissipative effects. The only variables at work are Θ and v_γ. In the Fourier space, the equation of continuity for Θ is

$$\frac{d\Theta}{dt} = -\frac{1}{3}kv_\gamma, \tag{8.9}$$

while the corresponding Euler equation is

$$\frac{dv_\gamma}{dt} = k\Theta, \tag{8.10}$$

where we have taken into account that the numerical density of photons n_γ scales as

$$n_\gamma \sim T^3, \tag{8.11}$$

and with the equation of state for photons:

$$p_\gamma \equiv \frac{\rho_\gamma}{3}, \tag{8.12}$$

where p_γ and γ_γ are respectively the pressure and the energy density of the photon fluid.

Combining the equation of continuity (8.9) with the Euler equation (8.10) in the Fourier space, one gets the harmonic oscillator equation:

$$\frac{d^2\Theta}{dt^2} + c_s^2 k^2 \Theta = 0, \tag{8.13}$$

where the sound speed is given by

$$c_s^2 = \frac{dp_\gamma}{d\rho_\gamma} \equiv \frac{1}{3}, \tag{8.14}$$

for the photon-dominated fluid. The solution of this oscillator equation can be written

$$\Theta(\eta) = \Theta(0)\cos(ks) + \frac{1}{kc_s}\frac{d\Theta}{dt}(0)\sin(ks), \tag{8.15}$$

where the value $\Theta(0)$ is relative to the initial conditions and s, such as in $ds = c_s \, d\eta$, is the horizon. In real space, these oscillations appear as standing waves for each Fourier mode. These standing waves continue to oscillate until recombination. At this point, the free-electron density drops drastically and the photons freely stream to the observer. The pattern of acoustic oscillations on the recombination surface seen by the observer becomes the acoustic peaks in the temperature anisotropy. Actually, assuming a negligible initial velocity perturbation, the temperature distribution at recombination η_{rec} is

$$\Theta(\eta_{rec}) = \Theta(0)\cos(ks_{rec}). \tag{8.16}$$

On small scales, the amplitude of the Fourier modes exhibits temporal oscillations as shown in Fig. 8.1. Modes caught in the extrema of their oscillations at recombination, corresponding to peaks in the spectrum, follow a harmonic relation:

$$k_n = \frac{\pi}{s_{rec}}n \quad \text{for } n = 1, 2, 3, \ldots. \tag{8.17}$$

As shown by Hu and White [205], observational verification of this harmonic series is the primary evidence for inflationary initial conditions.

A spatial perturbation in the CMB temperature of wavelength λ appears as an angular anisotropy of scale $\theta \sim \lambda/D$, where $D(z)$ is the comoving angular diameter distance to redshift z. The above harmonic relation implies a series of acoustic peaks in the anisotropy spectrum, located at

$$l_n = \frac{\pi D_{rec}}{s_{rec}}n \quad \text{for } n = 1, 2, 3, \ldots. \tag{8.18}$$

In a flat ($D_{\text{rec}} = \eta_0 - \eta_{\text{rec}}$) matter-dominated ($\eta \equiv \sqrt{a}$) universe, we have

$$\frac{\eta_{\text{rec}}}{\eta_0} \sim 2° \quad \text{or} \quad l_1 \sim 200. \tag{8.19}$$

Let us only remark that in a spatially curved universe, the angular diameter distance no longer equals the coordinate distance. One can show that a given comoving scale at a fixed distance subtends a larger (smaller) angle in a closed (open) universe than a flat universe (see Fig. 8.2).

Therefore, the observed first peak at $l_1 \sim 200$ (see Fig. 8.1) constrains the universe to be nearly spatially flat. Note that, for any perturbation in the primordial plasma before recombination, even this simple approximation of a photon-dominated perfect fluid already leads to the above description of acoustic oscillations in the temperature power spectrum. Thus far, we have neglected gravitational forces and redshifts in our discussion of our plasma description. Now, we shall include these effects.

8.3.2 Oscillations in the photon-dominated fluid with spatial curvature Φ and gravitational potential Ψ

Here, we shall see how the gravitational potential perturbations from inflation can be the sources of the initial temperature fluctuations. For a flat cosmology, one can write the metric

$$ds^2 = a^2\left[(1 + 2\Psi)\,d\eta^2 - (1 + 2\Phi)\,dx^2\right]. \tag{8.20}$$

The continuity equation, (8.9), for the fluctuations in the Fourier space becomes

$$\frac{d\Theta}{dt} = -\frac{1}{3}kv_\gamma - \frac{d\Psi}{dt}, \tag{8.21}$$

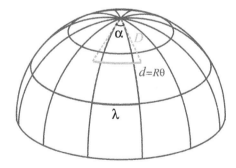

Fig. 8.2 In a closed universe, objects are further than they appear to be from Euclidean (flat) expectations corresponding to the difference between coordinate distance d and angular distance D. Consequently, at a fixed coordinate distance, a given angle corresponds to a smaller spatial scale in a closed universe. Acoustic peaks therefore appear at larger angles or lower l in a closed universe. The converse is true for an open universe (illustration from Hu and Dodelson [211])

while the corresponding Euler equation (8.10) can be written

$$\frac{dv_\gamma}{dt} = k(\Theta + \Psi). \tag{8.22}$$

Combining these two last equations one obtains the *forced harmonic oscillator* equation:

$$\frac{d^2\Theta}{dt^2} + c_s^2 k^2 \Theta = -\frac{k^2}{3}\Psi - \frac{d^2\Phi}{dt^2}. \tag{8.23}$$

To better come to know the effect of gravity on acoustic oscillations we need now to specify Φ and Ψ through a cosmological Poisson equation. Here, we only consider the particular case of constant gravitational potentials. The oscillator equation (8.23) becomes

$$\frac{d^2\Theta}{dt^2} + \frac{d^2\Psi}{dt^2} + c_s^2 k^2(\Theta + \Psi) = 0. \tag{8.24}$$

The solution given for adiabatic initial conditions in (8.16) can be generalized by

$$[\Theta + \Psi](\eta) = [\Theta + \Psi](0)\cos(ks). \tag{8.25}$$

Like a mass on a spring in a constant gravitational field of the earth, the solution represents oscillations around a displaced minimum. Furthermore, $\Theta + \Psi$ is also the observed temperature fluctuation. Photons lose energy climbing out of gravitational potentials at recombination and so the observer at present will see

$$\frac{\Delta T}{T} = \Theta + \Psi. \tag{8.26}$$

For the observer, acoustic oscillations are unchanged, but now he observes initial perturbations in the effective temperature which oscillate around zero with a frequency given by the sound speed. In fact, what the consideration of gravity adds is a way of connecting the initial conditions to inflationary curvature fluctuations.

8.3.3 Oscillations in the photon-dominated fluid with Θ, Ψ and baryonic effects: baryon loading

Now, we consider that baryons add extra mass to the above tightly coupled plasma with gravity. If the continuity equation remains given by

$$\frac{d\Theta}{dt} = -\frac{1}{3}kv_\gamma - \frac{d\Phi}{dt}, \tag{8.27}$$

one can show that the Euler equation is

$$\frac{d[(1 + R)v_\gamma]}{dt} = k\Theta + (1 + R)k\Psi, \tag{8.28}$$

with

$$R \equiv \frac{p_b + \rho_b}{p_\gamma + \rho_\gamma}. \qquad (8.29)$$

Therefore, the oscillator equation becomes

$$\frac{d}{dt}\left((1+R)\frac{d\Theta}{dt}\right) + \frac{k^2}{3}\Theta$$
$$= -\frac{k^2}{3}(1+R)\Psi - \frac{d}{dt}\left((1+R)\frac{d\Phi}{dt}\right). \qquad (8.30)$$

One can show that after several simplifications

$$\Phi = -\Psi = \text{constant} \quad \text{and} \quad \frac{1}{R}\frac{dR}{dt} \ll \omega,$$

where ω is the frequency of oscillations, and the above solution is modified as follows:

$$[\Theta + (1+R)\Psi](\eta) = [\Theta + (1+R)\Psi](0)\cos(ks). \qquad (8.31)$$

These equations are those of a mass $m \equiv 1 + R$ added to the spring in a constant gravitational field. One can show that there are several baryonic effects.

- The amplitude of the oscillations increases by a factor of $(1 + 3R)$.
- The *zero point* of the oscillations is shifted.
- The symmetry of the oscillations is broken so that even and odd peaks have different amplitudes.

Therefore, the modulation of the peak heights provides most of the information about the baryon–photon ratio in the acoustic peaks. One can say that, in particular, with the precise measurement of the first and second peaks, one obtains for the baryon–photon ratio a better sensitive value than the value given by nucleosynthesis.

8.3.4 Oscillations in the photon-dominated fluid with radiation effects: radiation driving

For all the above acoustic oscillations and their peaks, we assumed that recombination roughly takes place, for reasonable values of cosmological parameters, at the epoch of energy-density equality (which corresponds to the redshift z_{eq}) between matter and radiation, i.e. when

$$\frac{\rho_m}{\rho_r} \sim 24 \frac{\Omega_m h^2}{(1+z_{eq})^3} \sim 1. \qquad (8.32)$$

Now, we consider peaks which correspond to wave numbers that began oscillating earlier than that epoch. These higher peaks must carry all the effects of the radiation-domination epoch.

Hu and Sugiyama [221] have shown that radiation has the unique effect of driving the acoustic oscillations by making

the gravitational force evolve with time while matter does not. They have also calculated [222] that across the horizon scale at matter–radiation equality the amplitude of these acoustic oscillations increases by a factor of 5 for a pure photon fluid and dark-matter universe and a factor of 4 when including neutrinos and baryons.

Therefore, observations of these high peaks bring about important information about the contents of the universe at recombination.

8.3.5 Oscillations in the two-fluid primordial system: photon and baryon damping

Until this point, photons and baryons are considered as a single perfect fluid, the so-called photon-dominated fluid approximation or *tight-coupling approximation* with, in particular, a single bulk velocity $v_b = v_\gamma$ and hence no entropy generation.

First, fluid imperfections are only associated with the Compton mean free path λ_C given by the relation

$$\lambda_C = \frac{1}{n e \sigma_T a_{rec}} \sim 2.5 \, \text{Mpc}. \qquad (8.33)$$

Damping can be thought of as the result of the random walk in the baryons that takes photons from hot into cold regions and vice and versa. In particular, Silk [223] has shown that dissipation can become strong at the diffusion scale, the distance a photon can random walk in a given time η:

$$\lambda_D = \lambda_C \sqrt{\eta \lambda_C}. \qquad (8.34)$$

Then, to consider these dissipative effects, one must treat the photons and baryons as separate systems and introduce two continuity equations:

$$\frac{d\Theta}{dt} = -\frac{1}{3}k v_\gamma - \frac{d\Phi}{dt}, \qquad (8.35)$$

$$\frac{d\delta_b}{dt} = -k v_b - 3\frac{d\Phi}{dt}, \qquad (8.36)$$

with two coupled Euler equations:

$$\frac{dv_\gamma}{dt} = k(\Theta + \Psi) - \frac{k}{6}\pi_\gamma - \frac{d\tau}{dt}(v_\gamma - v_b), \qquad (8.37)$$

$$\frac{dv_b}{dt} = -\frac{1}{a}\frac{da}{dt}v_b + k\Psi + \frac{d\tau}{dt}\frac{v_\gamma - v_b}{R}, \qquad (8.38)$$

where δ_b and v_b are respectively the density perturbation and the bulk velocity of the baryons and where

$$R = \frac{p_b + \rho_b}{p_\gamma + \rho_\gamma} \sim 3\frac{\rho_b}{\rho_\gamma} \qquad (8.39)$$

is the photon–baryon momentum-density ratio. Note that the third term in both Euler equations—with $d\tau/dt$ given by

(8.33)—is a momentum-exchange term from Compton scattering but with opposite signs, since the total momentum is conserved. The second term in the Euler equation of the photon is due to radiation viscosity π_γ, which can be given by

$$\pi_\gamma = 2k v_\gamma \left(\frac{d\tau}{dt}\right)^{-1} A_\gamma, \tag{8.40}$$

with $A_\gamma = 16/15$ [212].

The first term in the Euler equation of the baryon is, of course, due to the cosmological expansion. Thus, the final *oscillator equation* becomes

$$c_s^2 \frac{d}{d\eta}\left(\frac{1}{c_s^2}\frac{d\Theta}{dt}\right) + k^2 c_s^2 \left(\frac{d\tau}{dt}\right)^{-1}[A_\gamma + A_h]\frac{d\Theta}{dt} + k^2 c_s^2 \Theta$$
$$= -\frac{k^2}{3}\Psi - c_s^2\frac{d}{d\eta}\left(\frac{1}{c_s^2}\frac{d\Theta}{dt}\right), \tag{8.41}$$

where A_h represents heat conduction with

$$A_h = \frac{R^2}{1+R}. \tag{8.42}$$

As for a mechanical oscillator, a term that depends on $d\Theta/dt$ provides a dissipation term to the solutions. Furthermore, a precise examination of this damped oscillator equation [211, 214] shows that diffusion damps the oscillation amplitude by a factor of order α_τ such as

$$\alpha_\tau = \exp\left[-k^2\eta\left(\frac{d\tau}{dt}\right)^{-1}\right], \tag{8.43}$$

the damping scale being

$$k_d \sim \sqrt{\frac{1}{\eta}\frac{d\tau}{dt}}; \tag{8.44}$$

see Fig. 8.3, where we plotted a numerical solution of (8.36) and (8.37) with the various effects on acoustic oscillations described above: gravity, baryon loading, radiation driving and damping.

We have described the various effects which can be at the origin of the wave form of the temperature anisotropy spectrum: see, for instance, in Fig. 8.1 the WMAP five-year TT power spectrum. We shall present later (Sect. 10) what precise pieces of information—such as cosmological parameters—have been extracted from such a spectrum by the WMAP team. But now it is time to introduce polarization and its different power spectra.

8.4 Polarization anisotropies

The CMB polarization is an important quest of the observational cosmology, because this observation permits one to characterize the earliest photons from the universe. After a long list of ever-decreasing upper limits, detection of polarization was made in 2002 by the DASI team at the Legendre number $l \sim 500$ and confirmed with the CAPMAP (i.e. Cosmic Anisotropy Polarization MAP) instrument with the measurement of the E-mode and B-mode power spectra [216].

The existence of polarization is a robust prediction of the standard cosmological picture, so a precise measurement of the CMB polarization should come as a confirmation of the standard model. Nevertheless, polarization measurements represent an experimental challenge. The weakness of the polarization signal requires both a demanding instrumental sensitivity and attention focused to all sources of systematic error.

Polarization of the CMB is naturally created by the Thompson scattering of photons off electrons at the moment of decoupling. The cross section of radiation scattered by a charged particle of charge e and mass m is given by

$$\frac{d\sigma}{d\varpi} = \left(\frac{e^2}{m_e c^2}\right)^2 |\hat{\mathbf{e}}.\hat{\hat{\mathbf{e}}}|^2, \tag{8.45}$$

where $\hat{\mathbf{e}}$ and $\hat{\hat{\mathbf{e}}}$ are the unit vectors indicating the linear polarization of the incident and scattered photon. This equation predicts that only a quadrupole distribution of initially unpolarized radiation will generate a linearly scattered polarization.

Thus, during the epoch of recombination, the electrons can produce a net linear polarization because of local quadrupole moments generated by Doppler-shifted radiation. The Doppler shift is the result of local velocity fields in the baryon–photon plasma. The scattered radiation will be polarized parallel to the crest and perpendicular to the wave vector.

In general, when the scattered polarization direction is either parallel or perpendicular to the wave vector of the perturbation, the polarization pattern is referred to as an E-mode pattern. When the polarization direction is at 45° to the wave vector, the polarization pattern is called a B-mode.

The decomposition of a polarization field into a sum of E- and B-modes has become the standard of analyzing polarization maps. Note that a B-mode pattern cannot be created by density perturbations and its presence would be a distinctive signature of primordial gravitational waves from inflation [210].

In terms of a multipole decomposition of the radiation field into spherical harmonics, Y_{lm}, the five quadrupole moments are represented by $m = 0, \pm 1$ and ± 2. The orthogonality of the spherical harmonics guarantees that no other moment can generate polarization from Thomson scattering.

The state of polarization is described in terms of Stokes parameters, and it is convenient to describe the observables

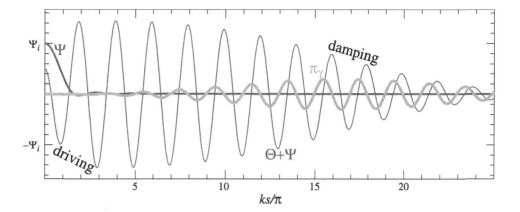

Fig. 8.3 Acoustic oscillations with gravitational forcing and dissipation damping; illustration from [213]

by a temperature fluctuation matrix decomposed in the Pauli basis [211]:

$$\mathbf{P} = \Theta(\check{\mathbf{n}})\mathbf{I} + Q(\check{\mathbf{n}})\sigma_3 + U(\check{\mathbf{n}})\sigma_1 + V(\check{\mathbf{n}})\sigma_2, \tag{8.46}$$

where I is the total intensity, Q the Stokes parameter of linear polarization, V for the circular polarization and U gives information on the phase. I and V remain invariant, but with a rotation angle ψ. Q and U are transformed thus:

$$\begin{pmatrix} Q' \\ U' \end{pmatrix} = \begin{pmatrix} \cos 2\psi & \sin 2\psi \\ -\sin 2\psi & \cos 2\psi \end{pmatrix} \begin{pmatrix} Q \\ U' \end{pmatrix}. \tag{8.47}$$

In cosmology the circular polarization V is absent; then we can decompose the field of polarization

$$E_{lm} \pm B_{lm} = -\int d\check{\mathbf{n}}_{\pm 2} Y^*_{lm}(\check{\mathbf{n}}) \big[Q(\check{\mathbf{n}}) \pm iU(\check{\mathbf{n}}) \big], \tag{8.48}$$

where the vectors of the fields are given by

$$E(\check{\mathbf{n}}) = \sum E_{lm} Y_{lm}(\check{\mathbf{n}}) \tag{8.49}$$

and

$$B(\check{\mathbf{n}}) = \sum B_{lm} Y_{lm}(\check{\mathbf{n}}). \tag{8.50}$$

In this case we can define the multiple moments which are used as the primary observable, i.e. the two-point correlation between the fields E and B, in order to define the polarization correlation power spectra $\Delta_{\mathrm{T}}^{(EE)}$ and $\Delta_{\mathrm{T}}^{(BB)}$:

$$\begin{aligned} \langle E^*_{lm} E_{l'm'} \rangle &= \delta_{ll'}\delta_{mm'} C_l^{(EE)}, \\ \langle B^*_{lm} B_{l'm'} \rangle &= \delta_{ll'}\delta_{mm'} C_l^{(BB)} \end{aligned} \tag{8.51}$$

and the temperature–polarization correlation power spectra

$$\langle \Theta^*_{lm} E_{l'm'} \rangle = \delta_{ll'}\delta_{mm'} C_l^{(TE)}. \tag{8.52}$$

The E-mode polarization power spectrum not only provides a more direct link to the properties of the last scattering

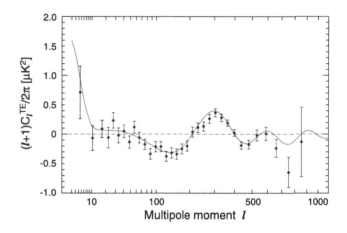

Fig. 8.4 The WMAP five-year TE power spectrum. The *green curve* is the best-fit theory spectrum from the ΛCDM/WMAP Markov chain [230]. The clear anticorrelation between the primordial plasma density (corresponding approximately to T) and velocity (corresponding approximately to E) in causally disconnected regions of the sky indicates that the primordial perturbations must have been on a superhorizon (curves from the CAPMAP experiment; see http://lambda.gsfc.nasa.gov/product/map)

surface than the temperature anisotropy but also offers complementary information which can be used to break various degeneracies in the determination of cosmological parameters. Nevertheless, the polarization signals (EE and BB) are much smaller than the temperature anisotropy spectrum; see Figs. 8.4, 8.5 and 8.6. Moreover, the polarized radiation is produced only near the end of recombination. The polarization spectra decline at large angular scales (low l), because photons could not diffuse so far before the end of recombination.

8.5 Secondary anisotropies

Beyond the peaks exits a lot of information on the origin and evolution of structure in the universe. As CMB photons travel from the recombination epoch to the observer, they traverse the large-scale structure of the universe and they pick up secondary temperature anisotropies. These sec-

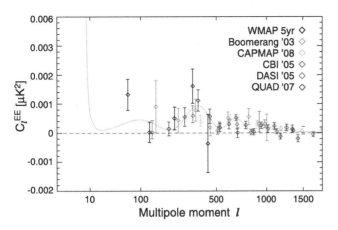

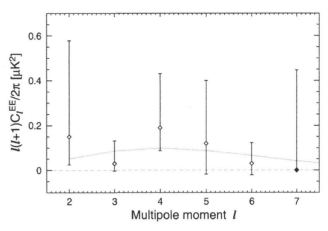

Fig. 8.5 WMAP five-year *EE* power spectrum, compared with results from the Boomerang (Montroy et al. [231], *green*), CBI (Sievers et al. [232], *red*), CAPMAP (Bischoff et al. [216], *orange*), QUAD (Ade et al. [233], *purple*), and DASI (Leitch et al. [234], *blue*) experiments. The *pink curve* is the best-fit theory spectrum from the ΛCDM/WMAP Markov chain [230], (curves from the CAPMAP experiment; see http://lambda.gsfc.nasa.gov/product/map and Nolta [235])

Fig. 8.6 WMAP five-year *EE* power spectrum at low *l*. The error bars are for 68% CL of the conditional likelihood of each multipole, with the other multipoles fixed at their fiducial theory values; the *diamonds* mark the peak of the conditional likelihood distribution. The error bars include noise and cosmic variance; the point at $l = 7$ is the 95% CL upper limit. The *pink curve* is the fiducial best-fit ΛCDM model [230] (curves from the CAPMAP experiment; see http://lambda.gsfc.nasa.gov/product/map)

ondary anisotropies depend on the temperature distribution and on the intervening dark matter, dark energy, baryonic density and, if they exist, on primordial gravitational waves. Secondary anisotropies can be divided in two classes: secondaries due to gravitational effects and secondaries induced by scattering off of electrons.

8.5.1 Gravitational secondaries anisotropy

There are two main sources for gravitational anisotropy: fluctuations of gravitational potential and gravitational lensing.

• *The ordinary Sachs–Wolfe effect* (SW effect) [196] is simply the gravitational redshift (or blueshift) due to the potential differences between the points of emission and reception of a photon.

• *The integrated Sachs–Wolfe effect* (ISW effect) comes when the gravitational potential changes with time along the photon trajectory: the photon accumulates a redshift as it travels, which translates into a temperature perturbation. The magnitude of this ISW effect can be written by an integral along the photon's path:

$$\left(\frac{\Delta T}{T}\right)_{ISW} = \int \left[\Psi^o(x, \eta) - \Phi^o(x, \eta)\right] d\eta, \qquad (8.53)$$

where, as seen above, Φ is the perturbation to the spatial curvature and Ψ is essentially the Newtonian gravitational potential. Let us remark that the ISW effect is linear in the perturbations. Finally, one can stress that the ISW effect is the most direct signature of the dark energy available in the CMB and is very sensitive to its equation of state.

• *The Rees–Sciama effect* (RS effect) [226]. When structure forms, linear perturbations become non-linear ones and the corresponding ISW effect may be called the RS effect. Note that the bulk motion of dark-matter halos of all masses contribute to the RS effect. But note also that this RS effect is much weaker than all the other effects.

• *The lensing effect.* The above gravitational secondaries (in particular the ISW effect) may be thought of as gravity imparting a "kick" to a photon forward or backward along the direction of motion. Gravity can also kick the photons in the transverse directions, changing their directions of motion but not their energies. The result of this weak gravitational lensing is that our image of the last scattering surface is slightly distorted. This results in a very slight smearing of the angular power spectrum, with power from the peaks being moved into the valleys [227]. Let us only note that gravitational lensing also generates a small amount of power in the anisotropies on its own. See all these gravitational anisotropies in Fig. 8.7 (left panel), where their amplitudes are plotted versus *l* or the angular scale $\theta = 180°/l$.

8.5.2 Scattering secondaries anisotropy

The main sources of scattering anisotropies are interactions of CMB photons with newly available electrons: the Sunyaev–Zel'dovich effect and reionization effects.

• *The Sunyaev–Zel'dovich effect* (SZ effect) [228]. CMB photons passing through clusters undergo Compton scattering by collisions with electrons in the ionized plasma causing spectral distortions in the CMB, where photons on the Rayleigh–Jeans side are transferred to the Wien tail. This largely measured effect gives important information on the

Fig. 8.7 Secondary anisotropies. (**a**) Gravitational anisotropies: ISW, lensing and Rees–Sciama (moving halo) effects. (**b**) Scattering secondaries: Doppler, density (δ) and ionization (i) modulated Doppler, and the SZ effect. Curves and model come from Hu and Dodelson [213]

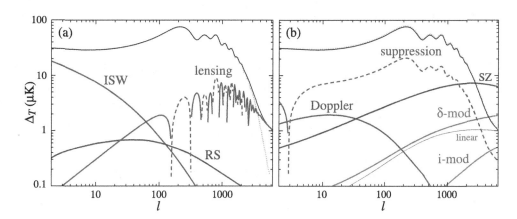

physics of clusters and on dark energy. Although this effect is supposed to dominate the power spectrum of secondary anisotropies, other effects of scattering anisotropies are measurable.

• *The reionization effects.* In all cosmological models, the universe is thought to have undergone reionization sometime between $6 < z < 30$: a few percent of the photons have rescattered since recombination. This rescattering induces: suppression of the peaks of the primary anisotropies by a $\exp(-\tau)$, where τ is the optical depth defined above; and the generation of polarization.

These effects are very interesting since they give information about the sources of ionizing radiation, i.e. about the epoch of the formation of the first objects in the universe. Note that the WMAP five-year data show, in particular, that the epoch of reionization is now more than 3σ earlier than $z = 6.2$. All of these scattering anisotropies are described in Fig. 8.7 (right panel), where their amplitudes are plotted versus l or the angular scale $\theta = 180°/l$.

As emphasized, in particular, by Melchiorri, one might add to this list: the *colored CMB anisotropies*. Actually, primary anisotropies can be partially erased at specific frequencies via Thomson resonant scattering, while the molecular clouds responsible for that are producing colored secondary anisotropies. Among the various possible primordial molecules, LiH has been chosen, in a first work [229], but the study is valid for other molecules, ions and atoms; see Sect. 9.

9 Cosmic microwave background anisotropies and primordial molecules

There have been several studies in order to gain understanding of chemical processes in the early universe as well as in the interactions of molecules with the CMB.

Dubrovich [236] analyzed the effect of hydrogen recombination lines on the CMB spectrum, while Rubiño–Martin et al. [237, 238] considered the effect of helium; see also the

recent discussions of Wong et al. [239] on the cosmological recombination process.

The main studies are relative to the postrecombination universe and possible imprints in the CMB from this period. The exhaustive work on the mechanisms of imprints from primordial molecules on the CMB was proposed by Schleicher et al. [167]. They presented the main effect of interaction and the rich physics which is involved.

Dubrovich [240, 241] analyzed the interaction of primordial molecules with the CMB, and the effect of various molecules due to their optical depths. Recently Mayer and Duschl [242] gave an overview of the opacity of the primordial gas. The main studies have focused on two directions:

• the influence on primary CMB fluctuations due to molecular optical depth, and
• the generation of secondary fluctuations due to scattering with protomolecular objects.

9.1 Imprint from primordial chemistry on CMB

The main mechanisms of interaction between CMB photons and primordial molecules are emission and absorption. Because their abundances are quite modest, collisional excitation and de-excitation are negligible and therefore emission and absorption are ineffective. The most obvious on the CMB is a frequency-dependent change in the observed CMB spectrum $I(\nu)$ due to absorption by a molecular species. For a given molecule, specified by the index *mol*, this effect is characterized by the attenuation, where $B(\nu)$ is the *incident* CMB spectrum:

$$I(\nu) = B(\nu)e^{-\tau_{\mathrm{mol}}}, \qquad (9.1)$$

where τ_{mol} is the optical depth of molecular species mol at the observed frequency ν_{obs}. If the emitted frequency is denoted ν_{em} at the redshift z_{em}, we have $\nu_{\mathrm{obs}} = (1 + z)\nu_{\mathrm{em}}$.

Thus along the line of sight the optical depth is expressed by

$$\tau_{\mathrm{mol}} = \int n_{\mathrm{mol}}\sigma(\nu)\,\mathrm{d}l, \qquad (9.2)$$

or with an integration over the redshift:

$$\tau_{\mathrm{mol}} = \int_0^{z_{\mathrm{em}}} cn_{\mathrm{mol}}(z)\sigma\left(\nu_{\mathrm{em}}(1+z)\right)\frac{\mathrm{d}z}{(1+z)H(z)}\,\mathrm{d}z, \quad (9.3)$$

where n_{mol} is the numerical density of the molecular species mol and $\sigma(\nu_{\mathrm{em}}(1+z))$ the absorption cross section of the considered species.

This optical depth can be caused by different processes of absorption, such as resonant line transitions, free–free processes or photodestruction of species which have a low photodissociation threshold.

Schleicher et al. [167] mentioned that this optical depth gives only an upper limit because inverse mechanisms, such as spontaneous and stimulated emission, can be balanced by the processes of absorption.

Thus, if we introduce the excitation temperature T_{ex}, which is determined by the ratio of collisional and radiative de-excitations, the frequency change in the radiation temperature T_{r} is expressed by [167]

$$\Delta T_{\mathrm{r}} = (T_{\mathrm{ex}} - T_{\mathrm{r}})\tau_{\mathrm{mol}}. \quad (9.4)$$

From this expression a few remarks can be made on the efficiency of the processes of absorption.

- Molecular species with high dipole moments provide a high cross section of absorption but have an excitation temperature close to T_{r}.
- Molecular species with low dipole moments give a low optical depth, but their excitation temperature is different from the CMB temperature.
- Photodestruction processes are governed by the radiation temperature, while the formation processes are regulated by the gas temperature, which is the same for free–free mechanisms.
- Molecular resonant line transitions can lead to a net change in the radiation temperature, as the photodestruction processes can also lead to a net change in the number of CMB photons. Thus, a few effects associated with the chemistry may affect the optical depth seen by CMB photons.

9.1.1 Molecular lines

The elastic scattering is a mechanism during which a photon is first absorbed and then re-emitted at the same frequency, but not in the same direction. One should note that for a given molecular line, ν_{ij}, each photon may be scattered only in a restricted range of redshift, when its redshifted frequency falls into the spectral width of the line such as

$$\frac{\Delta z}{z} \sim \frac{\Delta\nu_{ij}}{\nu_{ij}}. \quad (9.5)$$

The optical depth τ_{mol}^{ij} is given by

$$\tau_{\mathrm{mol}}^{ij} = \int cn_{ij}^{\mathrm{mol}}\sigma_d\frac{\mathrm{d}z}{(1+z)H(z)}, \quad (9.6)$$

where n_{ij}^{mol} is the numerical density of the molecule diffusing at the level ij. The cross section of elastic scattering is given by the Sobolev diffusion cross section:

$$\sigma_d = \frac{(\lambda_{ij}^{\mathrm{mol}})^3}{4c}A_{ij}\frac{\nu_{ij}}{\Delta\nu_{\mathrm{D}}}, \quad (9.7)$$

where λ_{ij} and ν_{ij} are respectively the wavelength and frequency of the transition roto-vibrational ij, A_{ij} is the first Einstein coefficient, and ν_{D} the Doppler broadening.

For a dipole moment of order of unity, Maoli et al. [229, 243] showed that the resonant scattering could be 10^{10}–10^{12} times more efficient than Thomson scattering (where $\sigma_{\mathrm{T}} = 6.652 \times 10^{-25}$ cm^2).

They therefore suggested why the coupling with primordial molecules could be stronger than the coupling with electrons in spite of a low abundance of primordial molecules. Thus, the resonant scattering being an elastic process, it can result in a possible CMB primary anisotropy attenuation.

Maoli et al. [229] discussed that the optical depth from primordial molecules may smear out primary fluctuations in the CMB. They showed that, if the angular scale θ is below the angular diameter of the horizon θ_{H}, any given power spectrum of primary CMB anisotropies could be affected.

Recently, Schleicher et al. [167] proposed calculations for the different optical depths (corrected for stimulated emission) for HeH$^+$ and HD$^+$, which are the most promising candidates. These molecules are interesting, because their dipole moment is relatively high, and because they are formed from quite abundant species. They showed that HeH$^+$ efficiently scatters CMB photons and smears out primordial fluctuations, leading to a change in the power spectrum close to 10^{-8}; see Fig. 9.1.

Notice that LiH is not considered because it has an even lower abundance [161] although the percentage of lithium converted to LiH is quite uncertain.

9.1.2 The negative hydrogen ion

Schleicher et al. [167] showed that there are two effects associated to the negative hydrogen ion H$^-$ that can significantly affect the optical depth seen by the CMB photons. The first one is relative to the bound–free process of photodetachment:

$$\mathrm{H}^- + \gamma \rightarrow \mathrm{H} + \mathrm{e}^-, \quad (9.8)$$

and the second one is relative to the free–free transitions that involve excited H$^-$:

$$\mathrm{H} + \mathrm{e}^- + \gamma \rightarrow [\mathrm{H}^-]^* \rightarrow \mathrm{H} + \mathrm{e}^-, \quad (9.9)$$

where $[H^-]^*$ is the intermediate state of excited H^-.

Schleicher et al. [167] showed that these are the most promising results. The free–free process leads to a frequency-change change in the power spectrum of the order 10^{-7} at 30 GHz; see Fig. 9.1.

9.1.3 Photodissociation of HeH⁺

Photodissociation of HeH^+ can give extra photons and can give distortion on the power spectrum. The resulting cross section was calculated by Dubrovich [143].

Schleicher et al. [167] showed that the absorption optical depth due to this process is between $[10^{11}, 5 \times 10^{-11}]$ for frequencies higher than 2 000 GHz; see Fig. 9.1.

9.2 On the primordial molecular clouds

From an initial idea of Zel'dovich, Dubrovich [240, 241] showed that resonant elastic scattering must be considered as one of the most efficient processes in coupling matter and radiation at high redshift. It was noted that the cross section for resonant scattering between photons and molecules is several orders of magnitude larger than that between radiation and electrons: even a modest abundance of primordial molecules would produce significant Thomson-type scattering.

At this point, every velocity field (due to molecular motion or to cloud infall) would leave its imprints on CMB via a Doppler shift. This technique for exploring the dark ages of the universe has been analyzed by de Bernardis et al. [244], Melchiorri and Melchiorri [245], Maoli et al. [229] and Signore et al. [246] and was summarized by Puy and Signore [247].

Secondary anisotropies are expected if the scattering centers have peculiar velocities. In this case the angular distribution of the scattered photons is no more isotropic in the CMB reference frame, due to the Mosengheil effect.

The intensity is given by [229]

$$\frac{\Delta I}{I} = (3 - \alpha)\beta \cos\theta \left(1 - \exp(-\tau)\right), \tag{9.10}$$

with

$$\alpha = \frac{\nu}{I} \frac{\partial I}{\partial \nu}. \tag{9.11}$$

The lithium hydride LiH molecule was considered as a very good candidate for resonant scattering with CMB photons, because the value of the dipole moment is high (i.e. 5.88 Debye). Maoli et al. [229] first considered the spectrum of a single cloud located at different redshifts but with a mass corresponding to a galaxy cluster today. Figure 9.2 showed the theoretical spectrum for a cloud at $z = 200$ and for lithium abundance $[LiH] = 3 \times 10^{-8}$.

In order to estimate the real level of anisotropy, one has to take into account the effect of other N clouds along the line of sight having lines redshifted at the same frequency.

Figure 9.3 shows the expected anisotropy versus the frequency of observation. The angular size of these anistropies would be that of protogalactic objects (i.e. $10''$–$100''$). We note that these secondary anisotropies have a maximum in the range 20–100 GHz. Of course ,the abundance of LiH in the molecular protoclouds are very high in comparison of the value given by the standard big-bang chemistry.

However, in the context of primordial molecular clouds, it is possible to think that the abundance could be higher, see the works of Puy and Pfenniger [152] on the differential chemistry in the baryonic fluctuations of matter.

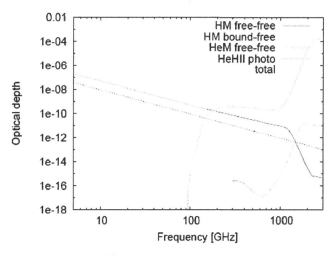

Fig. 9.1 The absorption optical depth due to different processes: HM line characterizes free–free processes, HM means bound–free process of H^- and the photodissociation of HeH^+ and the total line is the total optical depth. These curves come from the paper of Schleicher et al. [167]

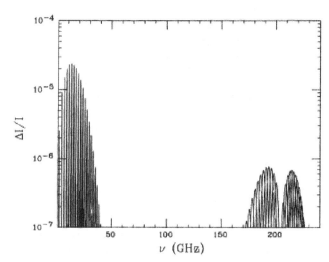

Fig. 9.2 Theoretical spectrum for a typical isolated cloud of galaxy cluster size at $z = 200$ moving at peculiar velocity. The abundance of LiH is close to $[LiH] = 3 \times 10^{-9}$ solar mass (curves from Maoli et al. [229])

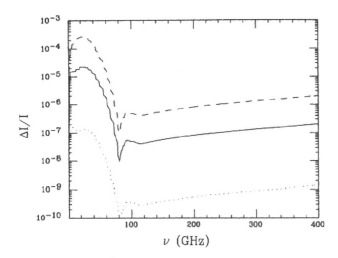

Fig. 9.3 Integrated theoretical spectrum (curves from Maoli et al. [229])

The search for primordial molecular lines may be one of the best tools to investigate the evolution of protostructure in the universe.

9.3 Observational situation

As said above, first searches for primordial resonant lines were carried out a few years ago using the 30 meter IRAM telescope at Pico Veleta (Spain) [244]. Observations were performed at three different frequencies: in the 1.3, 2, 3 mm atmospheric windows. They were characterized by a small band width and concerned only five spots in the sky. In this case, one obtained only upper limits for the abundance of the considered chemical species as a function of the redshift for a small interval in z and in a given direction.

A spectral line survey is performed using the ODIN satellite. This is a Swedish satellite developed in collaboration with France, Finland and Canada, equipped with a 1.1 meter telescope. Two sky regions—towards two of WMAP's hot spots [191]—are observed in the two frequency ranges of 542–547 and 486–492 GHz, with a resolution of 1 mK over 10 MHz and with a particular strategy for detecting lines produced by nearly formed structures.

The future for primordial resonant lines will be the ESA's HERSCHEL Space Observatory, which will allow one to observe a statistically representative area in the sky with a very large frequency range and a high sensitivity.

Primordial resonant lines are one of the main tools for exploring the pre-reionization and the post-reionization universe. The rich information associated with spectral lines can help to understand the formation and evolution of structures as well as the history of the metal enrichment of the high-redshift medium.

10 Past and future observations: COBE, WMAP, PLANCK

CMB is a cornerstone of modern cosmology . It is not surprising that the CMB provides us with a quantitative tool for studying the evolution of the universe. Because of atmospheric absorption, it was long ago realized that measurements of the high-frequency part of the CMB spectrum (wavelengths shorter than about 1 mm) should be performed from space. A satellite instrument also gives full sky coverage and a long observation time.

Observations by the NASA satellite the *Cosmic Background Explorer* (i.e. COBE) marked a turning point in CMB research both by establishing a definitive limit on spectral distortions through FIRAS measurements [119] and by discovering the temperature anisotropy through DMR measurements [12].

The NASA satellite the *Wilkinson Microwave Anisotropy Probe* (i.e. WMAP)—a descendant of the legendary COBE—has been mapping the temperature and polarization anisotropies of the CMB radiation. The PLANCK ESA's satellite is currently scheduled for launch at the beginning of 2009. In the following, we give a brief overview of these three spatial missions together with some long-duration ballooning observations.

10.1 The Cosmic Background Explorer COBE

In November 1989, NASA launched the COBE satellite. During the duration of its flight, the three instruments on board COBE (FIRAS, DIRBE and DMR) were mapping the sky at wavelengths from 1 µm to 1 cm with the fundamental missions of

- making precise measurements of the spectrum and of the large scale anisotropy of the cosmic microwave background back in redshift to $z \sim 10^6$ or $t \sim 1$ year;
- detecting a possible cosmic infrared background (hereafter CIB) due to the cumulative emissions of objects formed since the decoupling of matter and radiation ($z \sim 10^3$ or $t \sim 10^4$ years), and
- studying the large-scale infrared and submillimeter emissions of the galaxy.

The Far Infrared Absolute Spectrophotometer (i.e. FIRAS)—with Mather as Principal Investigator—covered wavelengths in two ranges: from 0.1 to 0.5 mm and from 0.5 to 1 mm. The purpose of the FIRAS was to compare the spectrum of the CMB radiation with that of an exact black body and therefore to measure the small deviations from a Planckian spectrum. The CMB spectrum—Fig. 4.2—and the definitive limits on both the Compton parameter y and the chemical potential μ were given by FIRAS [119] as seen in Sect. 4.

Fig. 10.1 False-color image of the near-infrared sky as seen by the DIRBE. Data at 1.25, 2.2, and 3.5 μm wavelengths are represented respectively as *blue, green* and *red colors*. The image is presented in galactic coordinates, with the plane of the Milky Way galaxy horizontal across the middle and the galactic center at the center. The dominant sources of light at these wavelengths are stars within our galaxy. The image shows both the thin disk and central bulge populations of stars in our spiral galaxy. Our sun, much closer to us than any other star, lies in the disk (which is why the disk appears edge-on to us) at a distance of about 28 000 light years from the center. The image is redder in directions where there is more dust between the stars absorbing starlight from distant stars. This absorption is so strong at visible wavelengths that the central part of the Milky Way cannot be seen. DIRBE data will facilitate studies of the content, energetics and large-scale structure of the galaxy, as well as the nature and distribution of dust within the solar system; figures from http://cmbdata.gsfc.nasa.gov/product/cobe/dirbeimage.cfm

The Diffuse Infrared Background Experiment (i.e. DIRBE)—with Hauser as Principal Investigator—spanned wavelengths from 1 to 300 μm in ten bands: J, K, L, M, the four IRAS bands (i.e. 12, 25, 60, 100 μm) and two large bands (100–200 μm and 200–300 μm). The cosmological objective of the DIRBE was to conduct a definitive search for CIB radiation. This purpose required extensive careful measuring and modeling of all the foreground emissions.

The DIRBE has mapped, in particular, the near-infrared sky: the false-color image of the near-infrared sky map is given in Fig. 10.1. The CIB had been first claimed by Puget et al. [248], using COBE data and was confirmed by several studies: those performed by Hauser et al. [249], Fixsen et al. [250], and Lagache et al. [251].

A set of Differential Microwave Radiometers (i.e. DMR) —with Smoot as Principal Investigator—operated at 3.3, 5.7 and 9.6 mm bands. The purpose of the DMR was to measure and map the large-scale anisotropy of the CMB radiation. The DMR results [12] are presented in Fig. 10.2, showing the near-uniformity of the CMB (top), the dipole (middle), and the multipole features of anisotropy (bottom).

Some cosmologists anticipated the 1992 COBE announcement. Almost at the same time that DMR measured these features, several observations were carried out by some nineteen other ground-based or balloon-borne experiments. But let us note that these anisotropies were not observed until one reached the full sky coverage and long integration times that were provided by the COBE space mission.

The Wilkinson Microwave Anisotropy Probe (i.e. WMAP) mission was the second satellite—devoted to CMB by NASA—which followed the COBE. The PLANCK surveyor will be the third generation of space missions—for the CMB experiment—but conducted by ESA, the European Space Agency.

To conclude on COBE, let us mention that Mather and Smoot have jointly received the Nobel Prize in Physics for 2006.

10.2 The balloon observations of millimetric extragalactic radiation and geophysics BOOMERANG

Waiting for the two planned major satellite experiments WMAP in 2001 and PLANCK Surveyor in 2009—each with full sky and wide frequency coverage—several balloon-borne experiments were developed at the end of the nineties. These smaller experiments covered smaller sky areas with comparable sensitivity and similar observational strategy

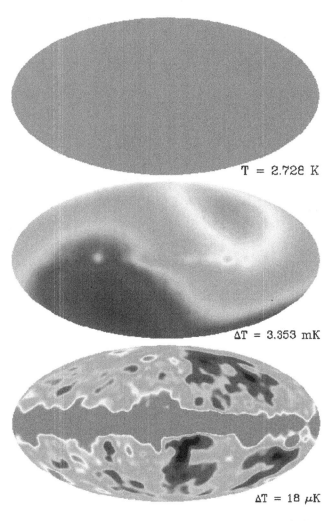

Fig. 10.2 The following image represents DMR data from the 53 GHz band (*top*) on a scale from 0–4 K, showing the near-uniformity of the CMB brightness (*middle*) on a scale intended to enhance the contrast due to the dipole described above, and (*bottom*) following subtraction of the dipole component; figures from http://cmbdata.gsfc.nasa.gov/product/cobe/dmrimage.cfm

compared to the two satellite missions. They used also the critical technologies which were under development for WMAP and PLANCK.

Let us present the most famous one: BOOMERANG (hereafter used for Balloon Observations Of Millimetric Extragalactic Radiation and Geophysics). This microwave telescope, from an idea of Melchiorri and Richards, was developed by the teams of de Bernardis and Lange. This instrument was intended to measure the CMB anisotropies at angular scales between a few degrees and ten arcminutes. It featured a wide focal plane with 16 detectors in the frequency bands centered at 90, 150, 220 and 400 GHz with FWHM ranging between 18 and 10 arcminutes. BOOMERANG was launched on 29 December 1998 from Mc Murdo Station in Antartica and was carried by a stratospheric balloon to an altitude of 38 km, where the low atmospheric pressure and low moisture content of the atmosphere facilitate such measurements.

The experiment collected data for 11 days without interruption, captured the first detection of the first peak due to acoustic oscillations localized at $l_{peak} = 197 \pm 6$ [207] and later provided the first high-resolution map of CMB anisotropies [252]. The teams upgraded the instrument and flew it again in 2003, this time detecting yet fainter patterns in the polarization of the CMB. Then, BOOMERANG provided the first high-resolution maps of CMB anisotropies. See, for instance, Fig. 10.3, for maps of CMB anisotropies and of the Stokes parameter, Q, as seen by the 2003 flight. Let us only remark that the color scaling of the Stokes parameter plot is identical to the temperature plot, emphasizing the tiny amplitude of the polarized signal.

In the previous section, we have seen that there is no doubt that we can see the acoustic wave projected on the sky. After the first release of the power spectra, for instance, Komatsu [260] asked *how the wave form in C_l determines the cosmological parameters.*

Fig. 10.3 *Left*: preliminary CMB anisotropy map in the region with deeper integration surveyed during the B2K flight. *Right*: map of the Stokes parameter Q obtained from the 8 PSB bolometers; curves from de Bernardis et al. [253]

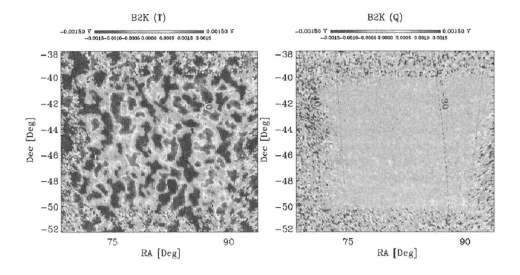

From the 2003 flight of BOOMERANG, they measured—not only the TT power spectrum [224]—but also the EE power spectrum [231]; moreover, they detected a significant TE correlation in the l-range between 50 and 950 [254]). It was the first measurement of the TE power spectrum using bolometer detectors.

Note that the TT and EE power spectra of the WMAP five-year data are compared with recent results of other experiments and in particular with the results [224, 231] obtained from the 2003 flight of BOOMERANG. Figures 8.1 and 8.5 show that now there exist many instruments devoted to CMB anisotropies and these moreover work with a great consistency between their various results. Before describing WMAP, let us mention that the Balzan Prize for 2006 was jointly awarded to de Bernardis and Lange.

10.3 The Wilkinson Microwave Anisotropy Probe WMAP

The WMAP satellite was launched in June 2001 from Cap Canaveral. It makes observations at the L2 Lagrangian point where the emission and magnetic field of the earth do not affect the satellite. The WMAP mission was designed to produce high-quality full-sky CMB maps, with the angular resolution a factor of 30 better than that of COBE, in five microwave bands: 22.8 GHz (or 13 mm), 33 GHz (or 9.1 mm), 40.7 GHz (or 7.3 mm), 60.8 GHz (or 4.9 mm), 93.5 GHz (or 3.2 mm).

WMAP released its first year results in February 2003. At that time, the mission was renamed the "Wilkinson Microwave Anisotropy Probe" to honor the late David Wilkinson who was a key "Investigator" of both COBE and WMAP missions until his death in September 2002.

On March 2008, WMAP released its five-year data set with new maps, power spectra and eight papers posted to LAMBDA; see http://lambda.gsfc.nasa.gov, [217, 230, 235, 255–258] and [259]). Figure 10.4 shows the five-year full-sky map of the CMB after foreground removal: colors represent the tiny temperature fluctuations of the *remnant glow* from the early universe.

From the only temperature TT power spectrum, in order to determine the parameters of a cosmological model, it is important to consider and to use all the features seen in Sect. 8.3—such as the locations, the heights of the peaks, etc.—which carry cosmological information.

- *Peak locations* lead to the distance to the last surface scattering which gives the 3D geometry, the age of the universe and the Hubble parameter altogether leading to the dark energy.
- *The first peak height* leads to the matter/radiation density ratio and to the dark-matter density.
- *The second peak height* leads to the baryon/radiation density ratio and therefore to the baryon density.

These parameters are primarily derived from the temperature data but the *fit* must include the polarization data which have a profound implication for cosmology.

(i) For instance, on possible reionization: from the C_l^{TE} released in 2003, WMAP was supposed to detect a significant TE correlation at $l < 7$ which gave a measure of the optical depth against Thomson scattering of $\tau = 0.17 \pm 0.04$ [261].

A naive interpretation of the 2003 WMAP polarization and quasar data was that the universe has begun to reionize at $z_r \sim 20$ and became completely reionized [260] by $z \sim 6$. As we shall see later, the new 2008 WMAP data favor a smaller optical depth and therefore a different conclusion for the possible reionization of the universe. But, anyway, τ is a primary key parameter.

(ii) On primordial fluctuations: the shape of the power spectrum of primordial curvature perturbations, $P_R(k)$, is one of the most powerful and practical tools for distinguishing among inflation models. In the ΛCDM

Fig. 10.4 Five-year WMAP

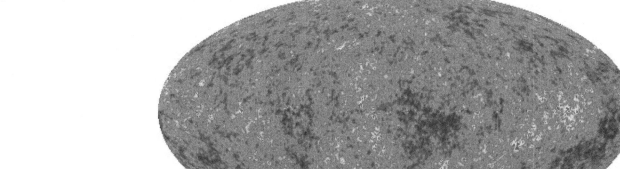

Table 10.1 Primary cosmological parameters of the ΛCDM model and their 68% intervals from *WMAP five-year mean*, adapted from [256]

$100\,\Omega_b h^2$	2.273 ± 0.062	for the parameter of baryon energy-density parameter
$\Omega_d h^2$	0.1099 ± 0.0062	for the parameter of the dark-matter energy density
Ω_Λ	0.742 ± 0.030	for the cosmological parameter of energy density
n_s	0.963 ± 0.014	for the spectrum index
τ	0.087 ± 0.017	for the optical depth against Thomson scattering
$\Delta_R^2(k)$	$2.410.11 \times 10^{-9}$	for discriminating between inflation models

model, primordial scalar fluctuations are adiabatic and Gaussian, and they can be described by

$$\Delta_R^2(k) = \frac{k^3}{2\pi^2} P_R(k) \quad \text{proportional to} \left(\frac{k}{k_0}\right)^{n_s - 1},$$

$$(10.1)$$

with $k_0 = 0.002\,\mathrm{Mpc}^{-1}$. $\Delta_R^2(k_0)$ and n_s are also considered as *key parameters*.

Now, let us note that the *wave form* is more and more accurately measured. Figures 8.1 and 8.4 show respectively the temperature TT and temperature–polarization correlation TE power spectra based on the five-year WMAP data. The EE power spectrum and other BB, TB and EB spectra are carefully studied [235]; see Fig. 8.5 for the EE power spectrum. Finally, Table 10.1 gives a set of six parameters of the minimal ΛCDM model and the corresponding 68% uncertainties derived from the five-year WMAP data.

Let us note that not only *primary parameters* but also *derived parameters* [256] are given. All these parameters are deduced from the WMAP five-year data but also when these data are combined with other data on Type-1a supernovae and on the distribution of the galaxies [256].

Komatsu et al. [256] have also explored the possible deviations from the simplest minimal ΛCDM model. These deviations appear as tests of inflation: flatness, adiabaticity and Gaussianity of the fluctuations, "scale invariance" of the power spectrum, and primordial gravitational waves. The WMAP team gives significant improvements on the limits on these deviations, while the satellite contributes to acquire more data. In conclusion, one can say that a lot of wonderful results can be found in the eight papers of the five-year WMAP data written by the members of its team.

Before describing the PLANCK mission, let us present parameters governing the sensitivity for both missions (see Table 10.2).

10.4 The future: the PLANCK satellite

10.4.1 On the satellite

PLANCK—see Fig. 10.6—is a mission of the European Space Agency designed to answer key cosmological ques-

Table 10.2 Parameters governing mission sensitivity from [263]. (a) NET/BLIP$_{CMB}$ is the ratio of detector sensitivity to the fundamental CMB background limit at 100 GHz, (b) NET/$1° \times 1°$ is the statistical noise equivalent CMB temperature at 100 GHz, assuming the total integration time to be spread evenly over the sky

Mission	WMAP (100 GHz)	PLANCK (100 GHz)
Duration	8 years	1.2 years
Detector number	8	8
NET/BLIP$_{CMB}$	~ 60	~ 4
NET/$1° \times 1°$	$= 6\,\mu K$	$= 1\,\mu K$

tions. It will be the first to map the entire sky with an unprecedented combination of sensitivity ($\Delta T/T \sim 2 \times 10^{-6}$), angular resolution (to $5'$) and frequency coverage (30–857 GHz). There are two instruments: the Low Frequency Instrument (i.e. LFI)—with HEMT arrays—covers the frequency range 30–70 GHz in three bands, and the High Frequency Instrument (i.e. HFI)—with bolometer arrays—covers the frequency range 100–857 GHz in six bands. Technical details of the PLANCK mission are given in *PLANCK: The Scientific Programme* [262]. Figure 10.5 shows the CMB spectrum and the nine frequency bands.

10.4.2 On the goals of PLANCK

Given that WMAP has already been extremely successful, it is natural to ask: what after WMAP?

Because WMAP can, crudely, measure less than 10% of the information contained in the CMB temperature anisotropies and only a tiny fraction of the information contained in the polarization anisotropies, PLANCK will do, a priori, a very good job. It can

- improve tests of inflationary models of the early universe,
- give accurate estimates of cosmological parameters and parameter degeneracies,
- give accurate estimates of the polarization power spectra,
- test non-Gaussianity,
- give significant new secondary science probes, in particular, as regards the Sunyaev–Zel'dovich effect,
- measure and identify foregrounds with astrophysics on these foregrounds.

Fig. 10.5 Spectra of sources in brightness temperature (in which a Rayleigh–Jeans ν^2 spectrum is flat), superimposed on the PLANCK frequency bands. Spectra of the galaxy and M82, a star-forming galaxy, are shown. For the galaxy, the components contributing to the over-all spectrum are identified. Also shown are the expected level of CMB fluctuations on a 1° scale, and, as a light dashed line (EG), the expected level of fluctuations introduced by all foreground radio sources on a 10′ scale; curves from the PLANCK science programme documents, ESA-SCI (2005)

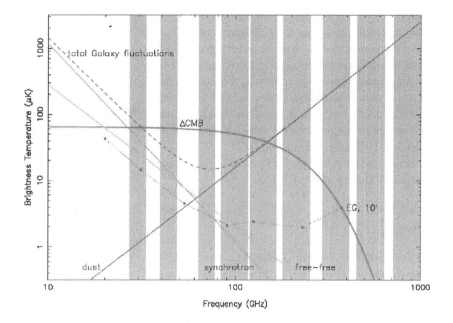

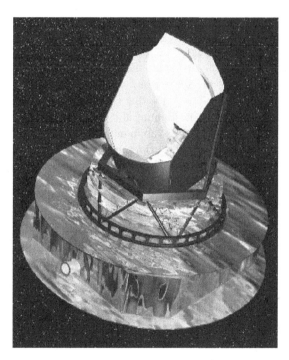

Fig. 10.6 PLANCK satellite, from the ESA web site www.esa.int

Fig. 10.7 HERSCHEL satellite, from the ESA web site www.esa.int

Concerning the effects of primordial chemistry on the CMB, Schleicher et al. [167] show that primordial chemistry does not alter the CMB at the significance level of the upcoming measurements by PLANCK. Nevertheless, the promising path to detect signals might be improved by the measurements of secondaries at small angular scales. These signals are related to the sizes of the primordial molecular clouds. This last information could be provided by the James Web Space Telescope (hereafter JWST) and the HER-SCHEL telescope.

10.4.3 Conclusions

The scientific return from PLANCK will be spectacular. With a great increase in capabilities over previous CMB missions, we wait for completely new science.

10.5 The HERSCHEL and ODIN satellite

Observational prospects for HERSCHEL and ODIN have been discussed by Maoli et al. [264]. The molecular lines produced by resonant scattering of the CMB photons are the most important signals coming from the dark age of the postrecombination universe. For these authors these signals associated with a linear evolution phase of perturbations are weaker and broader and their detection would be possible, as well as the primary anisotropy smearing, with millimetric multifrequency photometers under the condition of a very high sensitivity and a very precise characterization of foregrounds. This last requirement seems to be very challenging with the HERSCHEL satellite, see Fig. 10.7.

Let us notice that the relevance for PLANCK has recently been assessed by Dubrovich et al. [265].

11 Summary

One of the major discoveries in astrophysics was made in 1964: this was the first observation of the isotropic cosmic microwave background (CMB) radiation, by Penzias and Wilson, which has been of fundamental significance for our understanding of the development of the universe.

Since then, the FIRAS instrument, aboard the NASA satellite COBE, showed that the spectral distribution of the radiation is exactly that predicted for a black body at [119]: $T = 2.728 \pm 0.002$ K.

The DMR instrument, also aboard COBE, discovered temperature anisotropies, i.e. measured temperature fluctuations of about $\Delta T / T \sim 10^{-5}$. These can be interpreted as being due to a spatially varying gravitational potential at the time the CMB radiation was emitted. This happened at about $z \sim 1\,000$, which corresponds to some $300\,000$ years after the big bang.

Following COBE considerable progress has been made in higher-resolution measurements of the temperature anisotropy. With 45 times the sensitivity and 33 times the angular resolution of the COBE mission, WMAP was supposed to vastly extend our knowledge of cosmology: WMAP was supposed to measure the physics of the photon–baryon fluid at recombination and therefore to constrain models of structure formation, the geometry of the universe, and inflation.

After the release of the *One-year WMAP results*, the satellite is running extremely well with further improvements on anisotropy measurements. In particular, NASA, in 2008, presents the *Five-year WMAP observations*.

Meanwhile, HERSCHEL and PLANCK have been planned by ESA for a launch at the beginning of 2009. In these excellent perspectives, theoretical analyses of primordial molecules and CMB anisotropies with their profound implications on our knowledge of the physics of the early universe are actively studied in order to fully exploit the future results of HERSCHEL and PLANCK.

We should add that an important step for the big-bang theory was the discovery that light elements—hydrogen, helium-3, helium-4, lithium-7, deuterium—could be in relative abundances (once corrected for stellar nucleosynthesis)

as predicted by an early *hot phase*, the primordial nucleosynthesis. Therefore, this phase constituted, in a certain manner, the initial conditions for primordial chemistry.

The present status and prospects are as follows.

• Precise calculations give exact predictions for the fraction of each light element—D, ^{3}He, ^{4}He, ^{7}Li—that we should find from the early universe for any given value of η, where $\eta =$ baryon number/photon number of photons. If the observed relic abundances of D and ^{3}He are in agreement with their predicted abundances, the observed relic abundances of ^{4}He and ^{7}Li pose a challenge to the SBBN (standard big-bang nucleosynthesis).

• The establishment of a Planck spectrum for the CMB radiation as well as its possible distortions are well studied. FIRAS aboard the NASA satellite COBE has shown that the CMB radiation is a pure black body and has given the 95% limits for the Compton distortion: $y < 1.5 \times 10^{-5}$ and for the chemical potential distortion: $\mu < 9 \times 10^{-5}$. These upper limits on any deviation of the CMB from a black-body spectrum place strong constraints on the energy transfer between the CMB and matter which can be expressed as functions of the redshift z.

• During the dark ages, the period between the recombination epoch and the first stars, molecules form from neutral and charged atomic species and open new channels of radiative exchange. The primordial chemistries of hydrogen, helium, deuterium and lithium allow us to investigate the first phase of structure formation in the universe.

• The CMB is very nearly isotropic, but a *dipole anisotropy* has been found a long time ago. In particular, the dipole anisotropy measured by the NASA satellite WMAP, $\Delta T = 3.346 \pm 0.017$ mK, shows that the solar system barycenter is moving at a velocity of 368 ± 2 km/s relative to the observable universe.

• For the first time, DMR aboard the NASA satellite COBE detected temperature anisotropies of the CMB: spatial variations of order $\Delta T / T \sim 10^{-5}$ across $10°$–$90°$ on the sky. These 10^{-5} variations in temperature represent the direct imprint of initial gravitational potential perturbations on the CMB photons, called the Sachs–Wolf effect.

• On smaller angular scales ($\sim 1°$), the general concept of CMB anisotropy uses a reasonable approximation known as the *tight-coupling* approximation: before the moment of recombination, the baryon–photon fluid existed as a single entity and after, the baryons and photons are completely decoupled fluids. Let us remember that in a spherical harmonic expansion of the CMB temperature field, the angular power spectrum specifies the contributions to the fluctuations coming from different multipoles, each corresponding to the angular scale $\theta = \pi / l$. The theory predicts that the temperature power spectrum has a series of peaks and troughs. Gravitational, baryonic, matter–radiation, dissipation effects on these acoustic phenomena are well understood and, as seen later, confirmed by observations.

• The main source of polarization from recombination is associated with the acoustic peaks in temperature. The EB harmonic description allows a direct confrontation between predictions and observations of the various angular power spectra.

• The BOOMERANG Team reported in 2000 the first images of structures in CMB anisotropies over a significant part of the sky and derived the temperature angular power spectrum of the CMB, $\langle TT \rangle$, from $l = 50$ to 600 with a first peak at multipole $l_{pk} = 197 \pm 6$ (1σ error).

• The NASA satellite WMAP, launched in 2001, has been mapping the temperature and polarization anisotropies of the CMB radiation on the full sky in five microwave bands: 22, 30, 41, 60, 94 GHz. The unprecedented quality of the WMAP data demands careful and rigorous analysis for deriving constraints. Note that the effect of undesired foreground emission—synchrotron, free–free and dust—from our galaxy and extragalactic sources are minimized by observing in five frequencies. In March 2008, the Team presented new full-sky maps based on data from the first five years of the WMAP sky survey. The team also presented the temperature and polarization angular power spectra of the CMB derived from these first five years of WMAP data. With greater integration time they, in particular, accurately measure the first few peaks in the TT power spectrum. More generally, the spectra are in excellent agreement with the best-fit, minimal, six-parameter ΛCDM model.

• The lines produced by resonant scattering of the CMB photons are among *the most important* signals coming from the dark ages of the postrecombination universe. The future for primordial resonant lines is clearly represented by the satellite HERSCHEL, which will allow us to observe a statistically representative area in the sky with a large band width and high sensitivity.

Measurements of the CMB anisotropy made in the last five years have clearly moved the cosmology model into an era of parameter determination till the inflationary epoch. As said above, a simple minimal ΛCDM model fits the five-year map WMAP data. However, as Komatsu recently notes: *Many popular inflation models have been either ruled out, or being in danger* (Talk given at the Chalonge school at the Observatoire de Paris, July 2008).

Moreover, the discussed results depend on the study of perturbations that are still in the linear, small-amplitude regime. We have also seen that problems will come with the confrontation with the non-linearities and difficulties associated with hydrodynamics and full chemistry.

On the other hand, many theoretical studies of primordial molecules are elaborated and we hope to observe them and the *echography* of the early universe. Finally, we also expect to answer the following question: when were the first objects formed?

We think that many of these mysteries will be cleared up by the release of data from the satellite PLANCK/HERSCHEL due to be launched in 2009.

In this review, we have mentioned some of the outstanding problems of modern cosmology: the origin of the fluctuations, the inflationary picture, the nature of dark energy, the nature of dark matter. Particle physics combined with astrophysics can lead to a better understanding of these problems and to a beginning of their solution: the very large—with the future space and ground-based telescopes—and the very small—with the Large Hadron Collider LHC and the International Linear Collider ILC—come together in the picture we have of the early universe.

Let us only emphasize that when we consider this view we are not only talking about the universe when it was a second or a nanosecond old; we are considering densities and temperatures which exceed anything that is experienced in the laboratories.

There is no other way of studying the conditions of the early universe. But there is a lot of new physics and astrophysics to be understood.

Acknowledgements We would like to dedicate our review to the memory of Professor Francesco Melchiorri—who died in July 2005—who was a key astrophysicist in the understanding of many research fields on the cosmic microwave background. He was a key person in our scientific lives.

M.S. is grateful to Professor Pierre Encrenaz, who made her feel very welcome at his Laboratory of Observatoire de Paris and Ecole Normale Superieure, for always supporting her research in cosmology and for initiating, in France, the search for primordial molecules at the IRAM 30 m telescope and later with the ODIN satellite. Finally, M.S. appreciates valuable discussions with Bianca Melchiorri, Francoise Combes and Jean-Michel Lamarre.

D.P. acknowledges the PNCG (Programme National de Cosmologie et Galaxies) and the PNPS (Programme National de Physique Stellaire) for their strong financial assistance, as well as Daniel Pfenniger, Viktor Dubrovich and Patrick Vonlanthen for their helpful advice.

The authors would like to thank Dieter Haidt for a very careful reading of the manuscript and fruitful suggestions.

References

1. A. Einstein, Ann. Phys. **17**, 891 (1905)
2. A. Einstein, Ann. Phys. **49**, 769 (1916)
3. A. Friedmann, Z. Phys. **10**, 377 (1922)
4. G. Lemaitre, Phys. Rev. **25**, 903 (1925)
5. W. De Sitter, Mon. Not. R. Astron. Soc. **78**, 3 (1917)
6. E. Hubble, Proc. Natl. Acad. Sci. **15**, 168 (1929)
7. A. Penzias, R. Wilson, Astrophys. J. **142**, 419 (1965)
8. M. White, D. Scott, J. Silk, Annu. Rev. Astron. Astrophys. **32**, 319 (1994)
9. R. Partridge, *3K: Cosmic Microwave Background Radiation* (Cambridge University Press, Cambridge, 1995)
10. W. Hu, S. Dodelson, Annu. Rev. A **40**, 171 (2002)
11. D. Puy, M. Signore, New Astron. Rev. **46**, 29 (2002)
12. G.F. Smoot et al., Astrophys. J. Lett. **396**, L1 (1992)
13. C. Bennett et al., Astrophys. J. Suppl. **148**, 1 (2003)
14. D. Spergel et al., Astrophys. J. Suppl. **170**, 377 (2007)

15. PLANCK Collaboration, ESA-SCI **1** (2005)
16. E. Kolb, M. Turner, *Gravitation and Cosmology* (Addison-Wesley, Reading, 1990)
17. P.J.E. Peebles, *Principles of Physical Cosmology* (Princeton University Press, Princeton, 1993)
18. J.A. Peacock, *Cosmological Physics* (Cambridge University Press, Cambridge, 1998)
19. F. Melchiorri, B. Olivo-Melchiorri, M. Signore, Riv. Nuovo Cimento **26**, 1 (2003)
20. A. Guth, Phys. Rev. D **23**, 347 (1981)
21. A.A. Starobinsky, Phys. Lett. B **91**, 99 (1980)
22. A.D. Linde, Phys. Lett. B **108**, 389 (1982)
23. A. Albrecht, P.J. Steinhard, Phys. Rev. Lett. **48**, 1220 (1982)
24. A.D. Linde, Phys. Lett. B **129**, 177 (1983)
25. J. Bock et al., astro-ph/0604101
26. P.J.E. Peebles, Astrophys. J. **153**, 1 (1968)
27. S. Seager, D. Sasselov, D. Scott, Astrophys. J. **128**, 407 (2000)
28. L. Bergstrom, A. Goobar, *Cosmology and Particles Astrophysics* (Springer, Berlin, 2004)
29. P. Peter, J.P. Uzan, *Cosmologie Primordiale* (Berlin, 2005)
30. P. de Bernardis et al., Astrophys. J. **564**, 559 (2002)
31. A. Benoit et al., Astron. Astrophys. **399**, L19 (2003)
32. D. Spergel et al., Astrophys. J. Suppl. **148**, 175 (2003)
33. L. Page et al., Astrophys. J. Suppl. **148**, 233 (2003)
34. S. Perlmutter et al., Nature **391**, 51 (1998)
35. B. Schmidt et al., Astrophys. J. **507**, 46 (1998)
36. S. Weinberg, *Gravitation and Cosmology: Principles and Applications of General Relativity* (Wiley, New York, 1972)
37. S. Weinberg, Rev. Mod. Phys. **61**, 1 (1989)
38. N. Straumann, Eur. J. Phys. **20**, 419 (1999)
39. S. Carroll, astro-ph/0004075
40. S. Weinberg, astro-ph/0005265
41. P. Salati, *The Cosmic Microwave Background*, ed. by C. Lineweaver et al. NATO-ASI Series C, vol. 502 (1997), p. 365
42. D.S. Akerib, S.M. Carroll, M. Kamionkowski, S. Ritz, *Summer Study on the Future of Particle Physics*, ed. by N. Graf (Snowmass, 2001), p. 409
43. E.W. Kolb, hep-ph/9810362
44. J. Ellis, Phys. Scr. T **85**, 221 (2000)
45. R. Wagoner, Astrophys. J. **179**, 343 (1973)
46. J. Yang, M. Turner, G. Steigman, D. Schramm, K. Olive, Astrophys. J. **281**, 493 (1984)
47. T. Walker, G. Steigman, D. Schramm, K. Olive, K. Kang, Astrophys. J. **376**, 51 (1991)
48. M. Smith, L. Kawano, R. Malaney, Astrophys. J. Suppl. **85**, 219 (1993)
49. S. Sarkar, Rep. Prog. Phys. **59**, 1493 (1996)
50. K. Olive, G. Steigman, T. Walker, Phys. Rep. **333**, 389 (2000)
51. S. Burles, K. Nollett, M. Turner, astro-ph/9903300
52. S. Burles, K. Nollett, M. Turner, Astrophys. J. **552**, L1 (2001)
53. B. Fields, S. Sarkar, Phys. Rev. D **66**, 010001 (2002)
54. H. Kurki-Suonio, Phys. Rev. D **37**, 2104 (1988)
55. R. Malaney, G. Mathews, Phys. Rep. **229**, 145 (1993)
56. K. Sumiyoshi, T. Kajino, C. Alcock, G. Mathews, Phys. Rev. D **42**, 3963 (1990)
57. G. Fuller, K. Jedamzik, G. Mathews, Phys. Lett. B **333**, 135 (1994)
58. A. Heckler, Phys. Rev. D **51**, 405 (1995)
59. K. Kainulainen, H. Kurki-Suonio, Sihvola. Phys. Rev. D **59**, 083585 (1999)
60. J. Lara, T. Kajino, G. Mathews, Phys. Rev. D **73**, 083501 (2006)
61. L. Kavano, fermilab-Pub-92/04
62. L. Kavano, http://www-thphys.physics.ox.ac.uk/user/SubirSarkar/bbn.html
63. K. Nollet, S. Burles, Phys. Rev. D **61**, 123505 (2000)
64. G. Steigman, Int. J. Mod. Phys. E **15**, 1 (2006)
65. D. Kirkman et al., Astrophys. J. Suppl. **149**, 1 (2003)
66. D. Spergel et al., Astrophys. J. Suppl. **170**, 377 (2007)
67. G. Steigman, Annu. Rev. Nucl. Part. Sci. **57**, 463 (2007)
68. Y.I. Izotov et al., Astrophys. J. Suppl. **108**, 1 (1997)
69. Y.I. Izotov et al., Astrophys. J. **500**, 188 (1998)
70. K. Olive, G. Steigman, Astrophys. J. Suppl. **97**, 49 (1995)
71. K. Olive et al., Astrophys. J. **483**, 788 (1997)
72. B. Fields, K. Olive, Astrophys. J. **506**, 177 (1998)
73. Y.I. Izotov, T.X. Thuan, Astrophys. J. **602**, 200 (2004)
74. K. Olive, E. Skillman, Astrophys. J. **617**, 29 (2004)
75. R. Gruenwald et al., Astrophys. J. **567**, 931 (2002)
76. M. Fukugita, M. Kawasaki, Astrophys. J. **646**, 1 (2006)
77. M. Peimbert et al., Astrophys. J. **666**, 636 (2007)
78. F. Spite, M. Spite, Astron. Astrophys. **115**, 357 (1982)
79. S.G. Ryan et al., Astrophys. J. **523**, 654 (1999)
80. S. Ryan et al., Astrophys. J. **530**, L57 (2000)
81. P. Bonifacio et al., Mon. Not. R. Astron. Soc. **292**, L1 (1997)
82. P. Bonifacio, P. Molaro, Mon. Not. R. Astron. Soc. **285**, 847 (1997)
83. T. Bania et al., Nature **415**, 54 (2002)
84. M. Tegmark et al., Phys. Rev. D **74**, 123507 (2006)
85. J.P. Kneller, G. Steigman, New J. Phys. **6**, 117 (2004)
86. G. Gamov, Phys. Rev. **70**, 572 (1946)
87. G. Gamov, Nature **162**, 680 (1948)
88. R.A. Alpher, R.C. Herman, Nature **162**, 774 (1948)
89. R.H. Dicke, P.J.E. Peebles, P.J. Roll, D.T. Wilkinson, Astrophys. J. **142**, 414 (1965)
90. R. Mandolesi, N. Vittorio, *The Cosmic Microwave Background: 25 Years Later* (Kluwer Academic, Dordrecht, 1990), p. 164
91. S.S. Holt, C.L. Bennett, V. Trimble, *After the First Three Minutes*. AIP Conference Proceedings, vol. 222 (1991)
92. M. Signore, C. Dupraz, *The Infrared and Submillimeter Sky after COBE*, NATO-ASI 502, 1992
93. B. Melchiorri, F. Melchiorri, Riv. Nuovo Cimento **17**, 1 (1994)
94. J.L. Sanz, E. Martinez-Gonzalez, L. Cayon, *Present and Future of the Cosmic Microwave Background*. Lecture Notes in Physics, vol. 429 (Springer, Berlin, 1994)
95. R.B. Partridge, *3K: The Cosmic Microwave Background Radiation* (Cambridge University Press, Cambridge, 1995)
96. C.H. Lineweaveret et al., *The Cosmic Microwave Background*, NATO-ASI 502, 1997
97. M. Signore, F. Melchiorri, *Topological Defects in Cosmology* (World Scientific, Singapore, 1998)
98. L. Maiani, F. Melchiorri, N. Vittorio, *3K Cosmology*. AIP Conference Proceedings, vol. 476 (1999)
99. F. Melchiorri, G. Sironi, M. Signore, New Astron. Rev. **43**(1999)
100. J. Bartlett, New Astron. Rev. **45**, 283 (2001)
101. M. Signore, A. Blanchard, New Astron. Rev. **45** (2001)
102. A. Blanchard, M. Signore, *Frontiers of Cosmology*. NATO-ASI, vol. 187 (2005)
103. F. Melchiorri, Y. Rephaeli, *Background Microwave Radiation* (Scuola Enrico Fermi, Societ Italiana di Fisica, 2005)
104. G. Gamov, Phys. Today **3**, 16 (1950)
105. G. Gamov, K. Danske, Vidensk. Seks. Mat.-Phys. Medd. **7**, 1 (1953)
106. G. Gamov, in *Astronomy*, ed. by A. Beers (Pergamon, New York, 1956)
107. R.A. Alpher, R.C. Herman, Phys. Rev. **75**, 1089 (1949)
108. R.A. Alpher, R.C. Herman, Rev. Mod. Phys. **22**, 153 (1950)
109. R.A. Alpher, R.C. Herman, Phys. Rev. **84**, 60 (1951)
110. R.A. Alpher, R.C. Herman, *Modern Cosmology in Retrospect*, ed. by B. Bertotti et al. (1990), p. 129
111. W.S. Adams, Astrophys. J. **93**, 11 (1941)
112. A. McKellar, Publ. Dom. Astrophys. Obs. **1**, 251 (1941)
113. G. Herzberg, *Diatomic Molecules* (Van Nostrand-Reinhold, New York, 1950)

114. P. Thaddeus, Annu. Rev. Astron. Astrophys. **10**, 305 (1972)
115. M.E. Kaiser, E.L. Wright, Astrophys. J. Lett. **356**, L1 (1990)
116. K.C. Roth, D.M. Meyer, I. Hawkins, Astrophys. J. Lett. **413**, L67 (1993)
117. G. Dall'Oglio et al., Phys. Rev. **13**, 1187 (1976)
118. G.F. Smoot, *The Cosmic Microwave Background*, ed. by C.H. Lineweaver et al. NATO ASI, vol. 502 (1997), p. 271
119. D.J. Fixsen et al., Astrophys. J. **473**, 576 (1996)
120. J.C. Mather et al., Astrophys. J. **420**, 440 (1994)
121. M.A. Janssen, *The Infrared and Submillimeter Sky after COBE*, ed. by M. Signore, C. Dupraz. NATO ASI, vol. 359 (1992), p. 391
122. R.A. Sunyaev, Y.B. Zel'dovich, Annu. Rev. Astron. Astrophys. **18**, 537 (1980)
123. L. Danese, G. De Zotti, Rev. Nuovo Cimento **7**, 277 (1977)
124. P. Salati, *The Infrared and Submillimeter Sky after COBE*, ed. by M. Signore, C. Dupraz. NATO ASI, vol. 359 (1992), p. 143
125. J. Ellis et al., Nucl. Phys. B **373**, 399 (1992)
126. M. Tegmark, J. Silk, A. Blanchard, Astrophys. J. **420**, 484 (1994)
127. A. Kogut et al., Astrophys. J. **419**, 1 (1993)
128. J.P. Ostriker, C. Thompson, Astrophys. J. **323**, L97 (1987)
129. J.J. Levin, K. Freese, D.N. Spergel, Astrophys. J. **389**, 44 (1992)
130. L. Danese, G. De Zotti, Astron. Astrophys. **84**, 364 (1980)
131. L. Danese, G. De Zotti, Astron. Astrophys. **107**, 39 (1982)
132. A.S. Kompaneets, Sov. Phys. JETP **4**, 730 (1957)
133. C. Burigana, L. Danese, G. De Zotti, Astron. Astrophys. **246**, 49 (1991)
134. C. Burigana, G. De Zotti, L. Danese, Astrophys. J. **379**, 1 (1991)
135. L. Danese, C. Burigana, in *Present and Future of the CMB*, ed. by J.L. Sanz et al.Lecture Notes in Physics, vol. 429 (Springer, Berlin, 1994), p. 28
136. A.P. Lightman, Astrophys. J. **244**, 392 (1981)
137. W. Hu, J. Silk, Phys. Rev. D **48**, 485 (1993)
138. P.J.E. Peebles, Astrophys. J. **153**, 1 (1968)
139. P.J.E. Peebles, R. Dicke, Astrophys. J. **154**, 891 (1968)
140. R. Weymann, Phys. Fluids **8**, 2112 (1965)
141. R. Weymann, Astrophys. J. **145**, 560 (1966)
142. I. Bernshtein, D. Bernshtein, V. Dubrovich, Astron. Z. **54**, 727 (1977)
143. V. Dubrovich, V. Stolyarov, Astron. Lett. **23**, 565 (1997)
144. V. Dubrovich, S.I. Grachev, Astrofizika **34**, 249 (1991)
145. S. Lepp, M. Shull, Astrophys. J. **280**, 465 (1984)
146. W. Latter, J. Black, Astrophys. J. **372**, 161 (1991)
147. D. Puy, G. Alecian, J. Le Bourlot, J. Léorat, G. Pineau des Forêts, Astron. Astrophys. **267**, 337 (1993)
148. D. Galli, F. Palla, Astron. Astrophys. **335**, 403 (1998)
149. P. Stancil, S. Lepp, A. Dalgarno, Astrophys. J. **509**, 1 (1998)
150. S. Lepp, P. Stancil, A. Dalgarno, J. Phys. B **35**, R57 (2002)
151. D. Pfenniger, D. Puy, Astron. Astrophys. **398**, 447 (2003)
152. D. Puy, D. Pfenniger, Astron. Astrophys. (2008 submitted)
153. W. Roberge, A. Dalgarno, Astrophys. J. **255**, 489 (1982)
154. W. Saslaw, D. Zipoy, Nature **201**, 767 (1967)
155. Y. Shchekinov, M. Entél, Sov. Astron. **27**(6), 622 (1983)
156. R. Bienek, A. Dalgarno, Astrophys. J. **228**, 635 (1979)
157. T. Oka, Phys. Rev. Lett. **45**, 531 (1980)
158. J. Tennyson, Rep. Prog. Phys. **57**, 421 (1995)
159. E. Herbst, Philos. Trans. R. Soc. Lond. A **358**, 2523 (2000)
160. E. Bougleux, D. Galli, Mon. Not. R. Astron. Soc. **288**, 638 (1997)
161. P. Stancil, S. Lepp, A. Dalgarno, Astrophys. J. **458**, 401 (1996)
162. F. Gianturco, P. Gori Giorgi, Phys. Rev. A **54**, 4073 (1996)
163. F. Gianturco, P. Gori Giorgi, Astrophys. J. **479**, 560 (1997)
164. A. Dickinson, F. Gadéa, Mon. Not. R. Astron. Soc. **318**, 2000 (1227)
165. D. Galli, F. Palla, astro-ph/0202329
166. T. Abel, P. Anninos, Y. Zhang, M. Norman, New Astron. **2**, 181 (1997)
167. D. Schleicher, D. Galli, F. Palla, M. Camenzind, R. Kleseen, M. Bartelmann, S. Glover, astro-ph/08033987
168. F. Hasenöhrl, Ann. Phys. **15**, 344 (1904)
169. F. Hasenöhrl, Ann. Phys. **16**, 589 (1905)
170. K. Mosengheil, Ann. Phys. **22**, 867 (1907)
171. B. Melchiorri, F. Melchiorri, M. Signore, New Astron. Rev. **46**, 603 (2002)
172. B. Partridge, D.T. Wilkinson, Phys. Rev. Lett. **18**, 557 (1967)
173. D. Sciama, Phys. Rev. Lett. **18**, 1065 (1967)
174. M. Rees, D. Sciama, Nature **213**, 374 (1967)
175. J.M. Stewart, D.W. Sciama, Nature **216**, 748 (1967)
176. C.V. Heer, R.H. Kohl, Phys. Rev. **174**, 1611 (1968)
177. P.J.E. Peebles, Wilkinson, Phys. Rev. **174**, 2168 (1971)
178. M. Forman, Planet Space Sci. **18**, 25 (1970)
179. J.A. Peacock, *Cosmological Physics* (Cambridge University Press, Cambridge, 1998)
180. E.K. Conklin, R.N. Bracewell, Nature **216**, 777 (1967)
181. R.N. Bracewell, E.K. Conklin, Nature **219**, 1343 (1968)
182. E.K. Conklin, R.N. Bracewell, Nature **222**, 971 (1969)
183. E.K. Conklin, IAU Symp. **44**, 518 (1972)
184. P.S. Henry, Nature **231**, 516 (1971)
185. B.E. Corey, D.T. Wilkinson, Bull. Am. Astron. Soc. **8**, 351 (1976)
186. G.F. Smoot, M.V. Gorenstein, R.A. Muller, Phys. Rev. Lett. **39**, 898 (1977)
187. R. Fabbri, J. Guidi, F. Melchiorri, V. Natale, Phys. Rev. Lett. **44**, 1563 (1980)
188. S.P. Boughn et al., Astrophys. J. **243**, L113 (1981)
189. P.M. Lubin et al., Astrophys. J. **298**, L1 (1985)
190. C.L. Bennet et al., Astrophys. J. **464**, L1 (1996)
191. C.L. Bennet et al., Astrophys. J. Suppl. **148**, 1 (2003)
192. T. Pamanabhan, *Theoretical Astrophysics III* (Cambridge University Press, Cambridge, 2002)
193. E. Branchini, M. Plionis, D.W. Sciama, Astrophys. J. **461**, L17 (1996)
194. A. Dekel, *Cosmic Flows 1999*, ed. by S. Courteau et al. ASP Conf. Ser., vol. 420 (2000)
195. J.C. Mather et al., Astrophys. J. **354**, 37 (1990)
196. R.K. Sachs, A.M. Wolf, Astrophys. J. **147**, L1 (1967)
197. F. Melchiorri et al., Astrophys. J. **250**, 1 (1981)
198. R.D. Davies, A.N. Lasenby, R.A. Watson, Nature **236**, 462 (1987)
199. P.M. Lubin, G.L. Epstein, G.F. Smoot, Phys. Rev. Lett. **50**, 616 (1983)
200. D.J. Fixsen, E.S. Cheng, D.T. Wilkinson, Phys. Lett. **50**, 620 (1983)
201. P.J.E. Peebles, J.T. Yu, Astrophys. J. **162**, 815 (1970)
202. A.G. Doroshkevich, Y.B. Zel'dovich, R.A. Sunyaev, Sov. Astron. **22**, 523 (1978)
203. J.R. Bond, G. Efstathiou, Astrophys. J. **285**, L45 (1984)
204. U. Seljak, Astrophys. J. **435**, 87 (1994)
205. W. Hu, M. White, Astrophys. J. **471**, 30 (1996)
206. W. Hu, N. Sugiyama, J. Silk, Nature **386**, 37 (1997)
207. P. de Bernardis et al., Nature **404**, 955 (2000)
208. S. Hanany et al., Astrophys. J. Lett. **545**, L5 (2000)
209. J.R. Bond, in *Cosmology and Large Scale Structure*, ed. by R. Schaeffer, J. Silk, M.J. Spiro, J. Zinn-Justin (Elsevier Science, Amsterdam, 1996), p. 469
210. M. Kamionkowski, A. Kosowski, Annu. Rev. Nucl. Part. Sci. **49**, 77 (1999)
211. W. Hu, S. Dodelson, Annu. Rev. Astron. Astrophys. **40**, 171 (2002)
212. N. Kaiser, Mon. Not. R. Astron. Soc. **202**, 1169 (1983)
213. W. Hu, Ann. Phys. **303**, 203 (2003)
214. W. Hu, astro-ph/08023688
215. C. Reichardt et al., astro-ph/08011491

216. C. Bischoff et al., astro-ph/08020888
217. G. Hinshaw et al., astro-ph/08030722
218. J.M. Bardeen, Phys. Rev. D **22**, 1882 (1980)
219. G. Efstathiou, *Physics of the Early Universe*. Proceedings of the Scottish Universities Summer School in Physics, vol. 36, p. 361
220. V.F. Mukhanovn, H.A. Feldmann, R.H. Brandenberger, Phys. Rep. **215**, 203 (1992)
221. W. Hu, Sugiyama, Astrophys. J. **1444**, 489 (1995)
222. W. Hu, Sugiyama, Astrophys. J. **1471**, 542 (1996)
223. J. Silk, Sugiyama, Astrophys. J. **1151**, 459 (1968)
224. W.C. Jones et al., Astrophys. J. **647**, 823 (2006)
225. A.C.S. Readhead et al., Astrophys. J. **609**, 498 (2006)
226. M. Rees, D. Sciama, Nature **217**, 511 (1968)
227. U. Seljak, Astrophys. J. **463**, 1 (1996)
228. R. Sunyaev, Y. Zel'dovich, Comments Astrophys. Space Phys. **4**, 173 (1972)
229. R. Maoli, F. Melchiorri, D. Tosti, Astrophys. J. **425**, 372 (1994)
230. J. Dunkley et al., astro-ph/08030586
231. T.E. Montroy et al., Astrophys. J. **647**, 813 (2006)
232. J.L. Sievers et al., Astrophys. J. **660**, 976 (2007)
233. P. Ade et al., Astrophys. J. **674**, 22 (2008)
234. E. Leitch et al., Astrophys. J. **624**, 10 (2005)
235. M.R. Nolta et al., astro-ph/08030593
236. V.K. Dubrovich, Pis'ma Astron. Z. **1**, 10 (1975)
237. J.A. Rubiño-Martin, J. Chluba, R.A. Sunyaev, Astron. Astrophys. **371**, 1939 (2006)
238. J.A. Rubiño-Martin, J. Chluba, R.A. Sunyaev, astro-ph/07110594
239. W.Y. Wong, D. Scott, Mon. Not. R. Astron. Soc. **375**, 1441 (2007)
240. V.K. Dubrovich, Sov. Astron. Lett. **3**, 128 (1977)
241. V.K. Dubrovich, Astron. Astrophys. Trans. **5**, 57 (1994)
242. M. Mayer, W.J. Duschl, Mon. Not. R. Astron. Soc. **358**, 614 (2005)
243. R. Maoli et al., Astrophys. J. **457**, 1 (1996)
244. P. de Bernardis et al., Astron. Astrophys. **269**, 1 (1993)
245. B. Melchiorri, F. Melchiorri, Riv. Nuovo Cimento **17**, 1 (1994)
246. M. Signore et al., Astrophys. J. Suppl. **92**, 535 (1994)
247. D. Puy, M. Signore, New Astron. Rev. **51**, 411 (2007)
248. J.L. Puget, A. Abergel, J.P. Bernard, Astron. Astrophys. **308**, L5 (1996)
249. M.G. Hauser et al., Astrophys. J. **508**, 25 (1998)
250. D.J. Fixsen et al., Astrophys. J. **508**, 123 (1998)
251. G. Lagache et al., Astron. Astrophys. **344**, 322 (1999)
252. P. de Bernardis et al., Astrophys. J. **564**, 559 (2002)
253. P. de Bernardis et al., astro-ph/0311396
254. F. Piacentini et al., Astrophys. J. **647**, 833 (2006)
255. R. Hill et al., astro-ph/08030570
256. E. Komatsu et al., astro-ph/08030547
257. M. Limon et al., Astrophys. J. Suppl. (2008 submitted)
258. B. Gold et al., astro-ph/08030715
259. E. Wright et al., astro-ph/08030577
260. E. Komatsu et al., *Background Microwave Radiation and Intra Cluster Cosmology*, ed. by F. Melchiorri, Y. Rephaeli. Course CLIX, Varenna, SIF (2005), p. 37
261. A. Kogut et al., Astrophys. J. Suppl. **148**, 161 (2003)
262. The Scientific Programme of PLANCK, ESA-SCI I, 2005. astro-ph/0604069,
263. J. Bock et al., astro-ph/0604101
264. R. Maoli et al., astro-ph/0411641
265. V.K. Dubrovich, *Exploring the Cosmic Frontier*. ESO Symposia (Springer, Berlin, 2007)

The 2009 world average of α_s

Siegfried Bethke[a]

MPI für Physik, Föhringer Ring 6, 80805 Munich, Germany

Received: 10 August 2009 / Published online: 23 October 2009

© Springer-Verlag / Società Italiana di Fisica 2009

Abstract Measurements of α_s, the coupling strength of the Strong Interaction between quarks and gluons, are summarised and an updated value of the world average of $\alpha_s(M_{Z^0})$ is derived. Special emphasis is laid on the most recent determinations of α_s. These are obtained from τ-decays, from global fits of electroweak precision data and from measurements of the proton structure function F_2, which are based on perturbative QCD calculations up to $\mathcal{O}(\alpha_s^4)$; from hadronic event shapes and jet production in e^+e^- annihilation, based on $\mathcal{O}(\alpha_s^3)$ QCD; from jet production in deep inelastic scattering and from Υ decays, based on $\mathcal{O}(\alpha_s^2)$ QCD; and from heavy quarkonia based on unquenched QCD lattice calculations. A pragmatic method is chosen to obtain the world average and an estimate of its overall uncertainty, resulting in

$$\alpha_s(M_{Z^0}) = 0.1184 \pm 0.0007.$$

The measured values of $\alpha_s(Q^2)$, covering energy scales from $Q \equiv M_\tau = 1.78$ GeV to 209 GeV, exactly follow the energy dependence predicted by QCD and therefore significantly test the concept of Asymptotic Freedom.

PACS 12.38.Qk

Contents

1 Introduction

Quantum Chromodynamics (QCD) is the gauge field theory of the Strong Interaction [1–4]. QCD describes the interaction of quarks through the exchange of massless vector gauge bosons, the gluons, using similar concepts as in Quantum Electrodynamics, QED. The underlying gauge structure is a SU(3) rather than the simple U(1) of QED, implying many analogies, but also basic new features. The carriers of the strong force are 8 massless gluons in analogy to the photon for the electromagnetic force. An important new aspect is that the gluons, carrying a new quantum number called colour, can interact with each other.

As a consequence of the gluon self-coupling, QCD implies that the coupling strength α_s, the analogue to the fine structure constant α in QED, becomes large at large distances or—equivalently—at low momentum transfers.[1] Therefore QCD provides a qualitative reason for the observation that quarks do not appear as free particles but only exist as bound states of quarks, forming hadrons like protons, neutrons and pions. Hadrons appear to be neutral w.r.t. the strong quantum charge.

The quark statistics of all known hadrons, their production cross sections and decay widths imply that there are

---

[a] e-mail: bethke@mppmu.mpg.de

[1] "Large" distances Δs correspond to $\Delta s > 1$ fm, "low" momentum transfers to $Q < 1$ GeV/c.

three different states of the strong charge. Quarks carry one out of three different colour charges, while hadrons are colourless bound states of 3 quarks or 3 antiquarks ("baryons"), or of a quark and an anti-quark ("mesons"). Gluons, in contrast to the electrically neutral photons, carry two colour charges.

QCD does not predict the actual *value* of α_s. For large momentum transfers Q, however, it determines the functional form of the *energy dependence* of α_s. While an increasingly large coupling at small energy scales leads to the "confinement" of quarks and gluons inside hadrons, the coupling becomes small at high-energy or short-distance reactions; quarks and gluons are said to be "asymptotically free", i.e. $\alpha_s \to 0$ for momentum transfers $Q \to \infty$.

The *value* of α_s, at a given energy or momentum transfer[2] Q, must be obtained from experiment. Determining α_s at a specific energy scale Q is therefore a fundamental measurement, to be compared with measurements of the electromagnetic coupling α, of the elementary electric charge, or of the gravitational constant. Testing QCD, however, requires the measurement of α_s over ranges of energy scales: one measurement fixes the free parameter, while the others test the specific QCD prediction of confinement and of asymptotic freedom.

In the regime of $\alpha_s(Q^2) \ll 1$, methods of perturbation theory are applied to predict cross sections and distributions of physical processes implying quarks and gluons in the initial, intermediate or final state. The non-perturbative region where α_s approaches or exceeds values of $\mathcal{O}(1)$ usually leads to methodological problems in the interpretation of measurements. Theoretical uncertainties therefore arise from the non-perturbative regime and from unknown higher-order terms of the perturbative expansion. These uncertainties, in most cases, can only be dealt with in rather pragmatic ways, and—with few exceptions—they dominate the errors of experimental determinations of α_s.

In this review the current status of measurements of α_s is summarised. Theoretical basics of QCD and of the predicted energy dependence of α_s are given in Sect. 2. Actual measurements of α_s are presented in Sect. 3. A global summary of these results and a determination of the world average value of $\alpha_s(M_{Z^0})$ are presented in Sect. 4. Section 5 concludes and gives an outlook to future requirements and developments.

2 Theoretical basics

The concepts of Quantum Chromodynamics are presented in a variety of text books and articles, as e.g. [5–10], so that

in the following, only a brief summary of the basics of perturbative QCD and the running coupling parameter α_s will be given.

2.1 Energy dependence of α_s

With the value of α_s known at a specific energy scale Q^2, its energy dependence is given by the renormalisation group equation

$$Q^2 \frac{\partial \alpha_s(Q^2)}{\partial Q^2} = \beta\big(\alpha_s(Q^2)\big). \tag{1}$$

The perturbative expansion of the β function is calculated to complete 4-loop approximation [11, 12]:

$$\beta\big(\alpha_s(Q^2)\big) = -\beta_0 \alpha_s^2(Q^2) - \beta_1 \alpha_s^3(Q^2)$$
$$- \beta_2 \alpha_s^4(Q^2) - \beta_3 \alpha_s^5(Q^2) + \mathcal{O}(\alpha_s^6), \tag{2}$$

where

$$\begin{aligned}
\beta_0 &= \frac{33 - 2N_f}{12\pi}, \\
\beta_1 &= \frac{153 - 19N_f}{24\pi^2}, \\
\beta_2 &= \frac{77139 - 15099N_f + 325N_f^2}{3456\pi^3}, \\
\beta_3 &\approx \frac{29243 - 6946.3N_f + 405.089N_f^2 + 1.49931N_f^3}{256\pi^4}
\end{aligned} \tag{3}$$

and N_f is the number of active quark flavours at the energy scale Q. The numerical constants in (3) are functions of the group constants $C_A = N$ and $C_F = (N^2 - 1)/2N$, for theories exhibiting SU(N) symmetry. For QCD and SU(3), $C_A = 3$ and $C_F = 4/3$.

A solution of equation 1 in 1-loop approximation, i.e. neglecting β_1 and higher-order terms, is

$$\alpha_s(Q^2) = \frac{\alpha_s(\mu^2)}{1 + \alpha_s(\mu^2)\beta_0 \ln \frac{Q^2}{\mu^2}}, \tag{4}$$

where μ^2 appears as an integration constant. Apart from giving a relation between the values of α_s at two different energy scales, μ^2 at which α_s is assumed to be known, and Q^2 being another scale for which α_s is being predicted, (4) also demonstrates the property of asymptotic freedom, i.e. $\alpha_s \to 0$ for $Q^2 \to \infty$, provided that $N_f < 17$.

Likewise, (4) indicates that $\alpha_s(Q^2)$ grows to large values and diverges to infinity at small Q^2: for instance, with $\alpha_s(\mu^2 \equiv M_{Z^0}^2) = 0.12$ and for typical values of $N_f = 2, \ldots, 5$, $\alpha_s(Q^2)$ exceeds unity for $Q \leq \mathcal{O}(100 \text{ MeV}{-}1 \text{ GeV})$. Clearly, this is the region where perturbative expansions in α_s are not meaningful anymore. Therefore energy

---

[2]Here and in the following, the speed of light and Planck's constant are set to unity, $c = \hbar = 1$, such that energies, momenta and masses are given in units of GeV.

scales below the order of 1 GeV are regarded as the non-perturbative region where confinement sets in, and where (1) and (4) cannot be applied.

Including β_1 and higher-order terms, similar but more complicated relations for $\alpha_s(Q^2)$, as a function of $\alpha_s(\mu^2)$ and of $\ln\frac{Q^2}{\mu^2}$ as in (4), emerge. They can be solved numerically, such that for a given value of $\alpha_s(\mu^2)$, choosing a suitable reference scale like the mass of the Z^0 boson, $\mu = M_{Z^0}$, $\alpha_s(Q^2)$ can be accurately determined at any energy scale $Q^2 \geq 1$ GeV^2.

With

$$\Lambda^2 = \frac{\mu^2}{e^{1/(\beta_0\alpha_s(\mu^2))}},$$

a dimensional parameter Λ is introduced such that (4) transforms into

$$\alpha_s(Q^2) = \frac{1}{\beta_0 \ln(Q^2/\Lambda^2)}. \tag{5}$$

Hence, the Λ parameter is technically identical to the energy scale Q where $\alpha_s(Q^2)$ diverges to infinity. To give a numerical example, $\Lambda \approx 0.1$ GeV for $\alpha_s(M_{Z^0} \equiv 91.2$ GeV$) = 0.12$ and $N_f = 5$.

In complete 4-loop approximation and using the Λ-parametrisation in the $\overline{\text{MS}}$ renormalisation scheme (see Sect. 2.4), the running coupling is given [13] by

$$
\begin{aligned}
\alpha_s(Q^2) = &\frac{1}{\beta_0 L} - \frac{1}{\beta_0^3 L^2}\beta_1 \ln L \\
&+ \frac{1}{\beta_0^3 L^3}\left(\frac{\beta_1^2}{\beta_0^2}\left(\ln^2 L - \ln L - 1\right) + \frac{\beta_2}{\beta_0}\right) \\
&+ \frac{1}{\beta_0^4 L^4}\left(\frac{\beta_1^3}{\beta_0^3}\left(-\ln^3 L + \frac{5}{2}\ln^2 L + 2\ln L - \frac{1}{2}\right)\right) \\
&- \frac{1}{\beta_0^4 L^4}\left(3\frac{\beta_1\beta_2}{\beta_0^2}\ln L + \frac{\beta_3}{2\beta_0}\right),
\end{aligned}
\tag{6}
$$

where $L = \ln(Q^2/\Lambda_{\overline{\text{MS}}}^2)$. The first line of (6) includes the 1- and the 2-loop coefficients, the second line is the 3-loop and the third and the fourth lines denote the 4-loop correction, respectively.

The functional form of $\alpha_s(Q)$, in 4-loop approximation and for 4 different values of $\Lambda_{\overline{\text{MS}}}$, is displayed in Fig. 1. The slope and dependence on the actual value of $\Lambda_{\overline{\text{MS}}}$ is especially pronounced at small Q^2, while at large Q^2 both the energy dependence and the dependence on $\Lambda_{\overline{\text{MS}}}$ becomes increasingly feeble.

The relative size of higher-order loop corrections and the degree of convergence of the perturbative expansion of α_s is demonstrated in Fig. 2, where the fractional difference in the energy dependence of α_s, $(\alpha_s^{(4\text{-loop})} - \alpha_s^{(n\text{-loop})})/\alpha_s^{(4\text{-loop})}$,

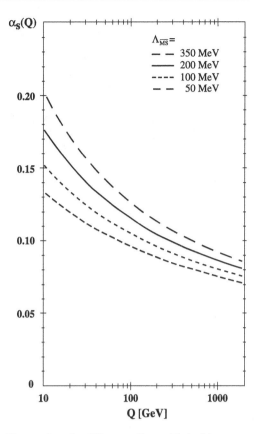

Fig. 1 The running of $\alpha_s(Q)$, according to (6), in 4-loop approximation, for different values of $\Lambda_{\overline{\text{MS}}}$

for $n = 1$, 2 and 3, is presented. The values of $\Lambda_{\overline{\text{MS}}}$ were chosen such that $\alpha_s(M_{Z^0}) = 0.1184$ in each order, i.e., $\Lambda_{\overline{\text{MS}}} = 90$ MeV (1-loop), $\Lambda_{\overline{\text{MS}}} = 231$ MeV (2-loop), and $\Lambda_{\overline{\text{MS}}} = 213$ MeV (3- and 4-loop). Only the 1-loop approximation shows sizeable differences of up to several per cent, in the energy and parameter range chosen, while the 2- and 3-loop approximations already reproduce the energy dependence of the 4-loop prediction quite accurately.

The parametrisation of the running coupling $\alpha_s(Q^2)$ with Λ instead of $\alpha_s(\mu^2)$ has become a common standard, see e.g. [9]. It will also be adopted here.

2.2 Quark threshold matching

Physical observables $\mathcal{R}$, when expressed as a function of α_s, must be continuous when crossing a quark threshold where N_f changes by one unit. This implies that Λ actually depends on the number of active quark flavours. Λ will therefore be labelled $\Lambda_{\overline{\text{MS}}}^{(N_f)}$ to indicate these peculiarities. Also the slope of the energy dependence and, in approximations higher than 2-loop, the value of α_s change at the quark flavour thresholds:

Construction of theoretical predictions which consistently match at a quark flavour threshold leads to matching conditions for the values of α_s above and below that

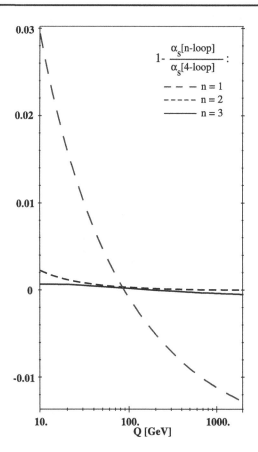

Fig. 2 Fractional difference between the 4-loop and the 1-, 2- and 3-loop presentations of $\alpha_s(Q)$, for $N_f = 5$ and $\Lambda_{\overline{MS}}$ chosen such that, in each order, $\alpha_s(M_{Z^0}) = 0.1184$

threshold [14]. In leading and in next-to-leading order, the matching condition is $\alpha_s^{(N_f-1)} = \alpha_s^{N_f}$. In higher orders, however, nontrivial matching conditions apply [13–15]. Formally these are of order $(n-1)$, if the energy evolution of α_s is performed in nth order (or n loops).

The matching scale $\mu^{(N_f)}$ can be chosen in terms of the (running) mass $m_q(\mu)$, or of the constant, so-called pole mass M_q. For both cases, the relevant matching conditions are given in [13]. These expressions have a particularly simple form for the choice[3] $\mu^{(N_f)} = m_q(m_q)$ or $\mu^{(N_f)} = M_q$. In this review, the latter choice will be used to perform 3-loop matching at the heavy quark pole masses, in which case the matching condition reads, with $a = \alpha_s^{(N_f)}/\pi$ and $a' = \alpha_s^{(N_f-1)}/\pi$:

$$\frac{a'}{a} = 1 + C_2 a^2 + C_3 a^3, \tag{7}$$

where $C_2 = -0.291667$ and $C_3 = -5.32389 + (N_f - 1) \cdot 0.26247$ [13].

---

[3]The results of reference [13] are also valid for other relations between $\mu^{(N_f)}$ and m_q or M_q, as e.g. $\mu^{(N_f)} = 2M_q$. For 3-loop matching, differences due to the freedom of this choice are negligible.

The fractional difference of the 4-loop prediction for the running α_s, using (6) with $\Lambda_{\overline{MS}}^{(N_f=5)} = 213$ MeV and 3-loop matching at the charm- and bottom-quark pole masses, $\mu_c^{(N_f=4)} = M_c = 1.5$ GeV and $\mu_b^{(N_f=5)} = M_b = 4.7$ GeV, and the 4-loop prediction without applying matching and with $N_f = 5$ throughout are illustrated in Fig. 3. Small discontinuities at the quark thresholds can be seen, such that $\alpha_s^{(N_f-1)} < \alpha_s^{(N_f)}$ by about 2 per mille at the bottom- and about 1 per cent at the charm-quark threshold. The corresponding values of $\Lambda_{\overline{MS}}$ are $\Lambda_{\overline{MS}}^{(N_f=4)} = 296$ MeV and $\Lambda_{\overline{MS}}^{(N_f=3)} = 338$ MeV. In addition to the discontinuities, the matched calculation shows a steeper rise towards smaller energies because of the larger values of $\Lambda_{\overline{MS}}^{(N_f=4)}$ and $\Lambda_{\overline{MS}}^{(N_f=3)}$. Note that the step function of α_s is not an effect which can be measured; the steps are artifacts of the truncated perturbation theory and the requirement that predictions for observables at energy scales around the matching point must be consistent and independent of the two possible choices of (neighbouring) values of N_f.

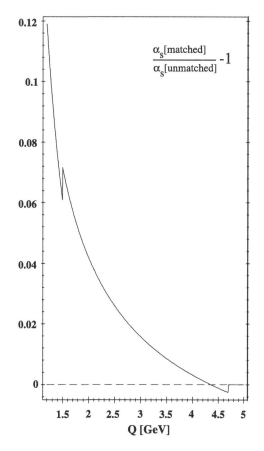

Fig. 3 The fractional difference between 4-loop running of $\alpha_s(Q)$ with 3-loop quark threshold matching according to (6) and (7), with $\Lambda_{\overline{MS}}^{(N_f=5)} = 213$ MeV and charm- and bottom-quark thresholds at the pole masses, $\mu_c^{(N_f=4)} \equiv M_c = 1.5$ GeV and $\mu_b^{(N_f=5)} \equiv M_b = 4.7$ GeV (*full line*), and the unmatched 4-loop result (*dashed line*)

2.3 Perturbative predictions of physical quantities

In perturbative QCD, physical quantities $\mathcal{R}$ are usually given by a power series in $\alpha_s(\mu^2)$, like

$$\mathcal{R}(Q^2) = P_l \sum_n R_n \alpha_s^n$$
$$= P_l\big(R_0 + R_1 \alpha_s(\mu^2)$$
$$+ R_2(Q^2/\mu^2)\alpha_s^2(\mu^2) + \cdots\big), \qquad (8)$$

where R_n are the nth order coefficients of the perturbation series and $P_l R_0$ denotes the lowest-order value of $\mathcal{R}$. R_1 is the *leading-order* (LO) or—equivalently—the first-order coefficient of the expansion in α_s, R_2 is called the *next-to-leading order* (NLO), R_3 is the *next-to-next-to-leading order* (NNLO) and R_4 the N3LO coefficient.[4]

QCD calculations in NLO perturbation theory are available for many observables $\mathcal{R}$ in high-energy particle reactions like hadronic event shapes, jet production rates, scaling violations of structure functions. Calculations including the complete NNLO are available for some totally inclusive quantities, like the total hadronic cross section in $e^+e^- \to$ hadrons, moments and sum rules of structure functions in deep inelastic scattering processes, the hadronic decay widths of the Z^0 boson and of the τ lepton. More recently, NNLO predictions were provided for exclusive quantities like hadronic event shape distributions and differential jet production rates in e^+e^- annihilation [16, 17], and N3LO predictions for the hadronic width of the Z^0 boson and the τ lepton [18] became available.

A further approach to calculating higher-order corrections is based on the resummation of logarithms which arise from soft and collinear singularities in gluon emission [19]. Application of resummation techniques and appropriate matching with fixed-order calculations are further detailed e.g. in [10].

2.4 Renormalisation

In quantum field theories like QCD and QED, physical quantities $\mathcal{R}$ can be expressed by a perturbation series in powers of the coupling parameter α_s or α, respectively. If these couplings are sufficiently small, i.e. if $\alpha_s \ll 1$, the series may converge sufficiently quickly such that it provides a realistic prediction of $\mathcal{R}$ even if only a limited number of perturbative orders will be known.

In QCD, examples of such quantities are cross sections, decay rates, jet production rates or hadronic event shapes.

Consider $\mathcal{R}$ being dimensionless and depending on α_s and on a single energy scale Q. When calculating $\mathcal{R}$ as a perturbation series in α_s, ultraviolet divergencies occur. These divergencies are removed by the "renormalisation" of a small set of physical parameters. Fixing these parameters at a given scale and absorbing this way the ultraviolet divergencies, introduces a second but artificial momentum or energy scale μ. As a consequence of this procedure, $\mathcal{R}$ and α_s become functions of the renormalisation scale μ. Since $\mathcal{R}$ is dimensionless, we assume that it only depends on the ratio Q^2/μ^2 and on the renormalised coupling $\alpha_s(\mu^2)$:

$$\mathcal{R} \equiv \mathcal{R}(Q^2/\mu^2, \alpha_s); \quad \alpha_s \equiv \alpha_s(\mu^2).$$

Because the choice of μ is arbitrary, however, the actual value of the experimental observable $\mathcal{R}$ cannot depend on μ, so that

$$\mu^2 \frac{\mathrm{d}}{\mathrm{d}\mu^2} \mathcal{R}(Q^2/\mu^2, \alpha_s)$$
$$= \left(\mu^2 \frac{\partial}{\partial\mu^2} + \mu^2 \frac{\partial\alpha_s}{\partial\mu^2} \frac{\partial}{\partial\alpha_s}\right) \mathcal{R} \overset{!}{=} 0, \qquad (9)$$

where the derivative is multiplied with μ^2 in order to keep the expression dimensionless. Equation (9) implies that any explicit dependence of $\mathcal{R}$ on μ must be cancelled by an appropriate μ-dependence of α_s to all orders. It would therefore be natural to identify the renormalisation scale with the physical energy scale of the process, $\mu^2 = Q^2$, eliminating the uncomfortable presence of a second and unspecified scale. In this case, α_s transforms to the "running coupling constant" $\alpha_s(Q^2)$, and the energy dependence of $\mathcal{R}$ enters only through the energy dependence of $\alpha_s(Q^2)$.

The principal independence of a physical observable $\mathcal{R}$ from the choice of the renormalisation scale μ was expressed in (9). Replacing α_s by $\alpha_s(\mu^2)$, using (1), and inserting the perturbative expansion of $\mathcal{R}$ (see (8)) into (9) results, for processes with constant P_l, in

$$0 = \mu^2 \frac{\partial R_0}{\partial\mu^2} + \alpha_s(\mu^2)\mu^2 \frac{\partial R_1}{\partial\mu^2} + \alpha_s^2(\mu^2)\left[\mu^2 \frac{\partial R_2}{\partial\mu^2} - R_1\beta_0\right]$$
$$+ \alpha_s^3(\mu^2)\left[\mu^2 \frac{\partial R_3}{\partial\mu^2} - [R_1\beta_1 + 2R_2\beta_0]\right]$$
$$+ \mathcal{O}(\alpha_s^4). \qquad (10)$$

Solving this relation requires that the coefficients of $\alpha_s^n(\mu^2)$ vanish for each order n. With an appropriate choice of inte-

---

[4]The notions of (next-to-) *leading order* and their mapping to nth order in α_s are not consistently treated in the literature. For instance, N3LO usually corresponds to $\mathcal{O}(\alpha_s^4)$ in e^+e^- annihilation [18], the notion N3LO is used for $\mathcal{O}(\alpha_s^3)$ in deep inelastic scattering, see e.g. [61].

gration limits one thus obtains

$$R_0 = \text{const.},$$

$$R_1 = \text{const.},$$

$$R_2\left(\frac{Q^2}{\mu^2}\right) = R_2(1) - \beta_0 R_1 \ln\frac{Q^2}{\mu^2},$$

$$R_3\left(\frac{Q^2}{\mu^2}\right) = R_3(1) - \left[2R_2(1)\beta_0 + R_1\beta_1\right]\ln\frac{Q^2}{\mu^2}$$
$$+ R_1\beta_0^2\ln^2\frac{Q^2}{\mu^2} \tag{11}$$

as a solution of (10).

Invariance of the complete perturbation series against the choice of the renormalisation scale μ^2 therefore implies that the coefficients R_n, except R_0 and R_1, explicitly depend on μ^2. In infinite order, the renormalisation scale dependence of α_s and of the coefficients R_n cancel; in any finite (truncated) order, however, the cancellation is not perfect, such that all realistic perturbative QCD predictions include a remaining explicit dependence on the choice of the renormalisation scale.

The scale dependence is most pronounced in leading order QCD because R_1 does not explicitly depend on μ and thus, there is no cancellation of the (logarithmic) scale dependence of $\alpha_s(\mu^2)$ at all. Only in next-to-leading and higher orders, the scale dependence of the coefficients R_n, for $n \geq 2$, partly cancels that of $\alpha_s(\mu^2)$. In general, the degree of cancellation improves with the inclusion of higher orders in the perturbation series of $\mathcal{R}$.

Renormalisation scale dependence is often used to test and specify uncertainties of theoretical calculations for physical observables. In most studies, the central value of $\alpha_s(\mu^2)$ is determined or taken for μ equalling the typical energy of the underlying scattering reaction, like e.g. $\mu^2 = E_{\text{cm}}^2$ in e^+e^- annihilation. Changes of the result when varying this definition of μ within "reasonable ranges" are taken as systematic higher-order uncertainties.

There are several proposals of how to optimise or fix the renormalisation scale; see e.g. [20–23]. Unfortunately, there is no common agreement of how to optimise the choice of scales or how to define the size of the corresponding uncertainties. This unfortunate situation should be kept in mind when comparing and summarising results from different analyses.

In next-to-leading order, variation of the renormalisation scale is sufficient to assess and include theoretical uncertainties due to the chosen renormalisation scheme and the limited (truncated) perturbation series. In NNLO and higher, however, *both* the renormalisation scale and the renormalisation scheme should be varied for a complete assessment. While it has become customary to include renormalisation scale variations when applying theoretical predic-

tions, changes of the renormalisation scheme are rarely explored. Instead, the so-called "modified minimal subtraction scheme" ($\overline{\text{MS}}$) [24] is commonly used in most analyses, which is also the standard choice in this review.

2.5 Non-perturbative methods

At large distances or low momentum transfers, α_s becomes large and application of perturbation theory becomes inappropriate. Non-perturbative methods have therefore been developed to quantify strong interaction processes at low-energy scales of typically $Q^2 < 1$ GeV^2, such as the fragmentation of quarks and gluons into hadrons ("hadronisation") and the masses and mass splittings of mesons.

Hadronisation models are used in Monte Carlo approaches to describe the transition of quarks and gluons into hadrons. They are based on QCD-inspired mechanisms like the "string fragmentation" [25, 26] or "cluster fragmentation" [27, 28], and are usually implemented, together with perturbative QCD shower and/or (N)LO QCD generators, in models describing complete hadronic final states in high-energy particle collisions. Those models contain a number of free parameters which must be adjusted in order to reproduce the experimental data well. They are indispensable tools not only for detailed QCD studies of high-energy collision reactions, but are also important to assess the resolution and acceptance of large particle detector systems.

Power corrections are an analytic approach to approximate non-perturbative hadronisation effects by means of perturbative methods, introducing a universal, non-perturbative parameter

$$\alpha_0(\mu_I) = \frac{1}{\mu_I}\int_0^{\mu_I} \mathrm{d}k\, \alpha_s(k)$$

to parametrise the unknown behaviour of $\alpha_s(Q)$ below a certain infrared matching scale μ_I [29–34]. Power corrections are regarded as an alternative approach to describe hadronisation effects on event shape distributions, instead of using phenomenological hadronisation models.

Lattice Gauge Theory is one of the most developed non-perturbative methods (see e.g. [35]) and is used to calculate, for instance, hadron masses, mass splittings and QCD matrix elements. In Lattice QCD, field operators are applied on a discrete, 4-dimensional Euclidean space-time of hypercubes with side length a. Finite size lattice and spacing effects are studied by using increased lattice sizes and decreased lattice spacing a, hoping to eventually approach the continuum limit. With ever increasing computing power and refined Monte Carlo methods, these calculations significantly matured over time and recently provided predictions of the proton (and other hadron) masses to better than 2% [36], and determinations of α_s from quarkonia mass splittings with a precision of better than 1% [37].

3 Measurements of α_s

Since almost 30 years, determinations of α_s continue to be at the forefront of experimental studies and tests of QCD. Increasing precision of QCD predictions and methods, improved understanding and parametrisation of non-perturbative effects, increased data quality and statistics and the availability of data over large ranges of energy and from a large variety of processes have led to an ever increasing precision and depth of these studies. The development of α_s determinations was documented and summarised in a number of summary articles, see e.g. [9, 10, 38, 39]. Since about the year 2000, the precision of α_s determinations and the multitude of results from various processes and ranges of energies provided experimental proof [10, 39] of the concept of asymptotic freedom.

This review aims at an update of the review from 2006 [39] which yielded a world average value of $\alpha_s(M_{Z^0}) = 0.1189 \pm 0.0010$. Here, special emphasis will be laid on the most recent results which are based on further improved theoretical predictions and/or experimental precision:

- perturbative QCD predictions in complete $\mathcal{O}(\alpha_s^4)$ (N3LO) for the hadronic widths of the Z^0 boson and the τ lepton are now available, improving further the completeness of the perturbative series and providing increased control of remaining theoretical uncertainties;
- improved lattice QCD simulations with vacuum polarisation from u, d and s quarks, updating previous determinations of α_s and quoting overall uncertainties of less than 1%;
- an improved extraction of α_s from radiative decays of the $\Upsilon(1s)$;
- a combined analysis of non-singlet structure functions from deep inelastic scattering data, based on QCD predictions complete to $\mathcal{O}(\alpha_s^3)$ (N3LO);
- a combined analysis of inclusive jet cross section measurements in neutral current deep inelastic scattering at high Q^2;
- determinations of α_s from hadronic event shapes and jet rates in e^+e^- annihilation final states, an important and (experimentally) very precise environment, based on the new and long awaited QCD predictions in complete NNLO QCD.

These recent results are superior to and thus supersede a large number of α_s determinations published before 2006 and summarised in [39].

3.1 α_s from τ-lepton decays

Determination of α_s from τ lepton decays is one of the most actively studied fields to measure this basic quantity. The small effective energy scale, $Q = M_\tau = 1.78$ GeV,

small non-perturbative contributions to experimental measurements of a total inclusive observable, the normalised hadronic branching fraction of τ lepton decays,

$$R_\tau = \frac{\Gamma(\tau^- \to \text{hadrons } \nu_\tau)}{\Gamma(\tau^- \to e^- \overline{\nu}_e \nu_\tau)}, \tag{12}$$

invariant mass distributions (spectral functions) of hadronic final states of τ-decays, and the "shrinking error" effect of the QCD energy evolution of α_s towards higher energies[5] provide the means for one of the most precise determinations of $\alpha_s(M_{Z^0})$. Theoretically, R_τ is predicted to be [40]

$$R_\tau = N_c S_{EW} |V_{ud}|^2 (1 + \delta'_{EW} + \delta_{\text{pert}} + \delta_{\text{nonpert}}). \tag{13}$$

Here, $S_{EW} = 1.0189(6)$ [41] and $\delta'_{EW} = 0.001(1)$ [42] are electroweak corrections, $|V_{ud}|^2 = 0.97418(27)$ [9], δ_{pert} and δ_{nonpert} are perturbative and non-perturbative QCD corrections. Most recently, δ_{pert} was calculated to complete N3LO perturbative order, $\mathcal{O}(\alpha_s^4)$ [18]; it is of similar structure as the one for the hadronic branching fraction R_Z of the Z^0 boson. Based on the operator product expansion (OPE) [43], the non-perturbative corrections are estimated to be small [40], $\delta_{\text{nonpert}} \sim -0.007 \pm 0.004$. A comprehensive review of the physics of hadronic τ decays was given in [44].

Since 2006, several authors have revisited the determination of α_s from τ decays [18, 45–50]. These studies are based on data from LEP [51, 52] and—partly—from BABAR [53]. They differ, however, in the detailed treatment and usage of the perturbative QCD expansion of R_τ. In particular, the usage of either a fixed-order (FOPT) or contour improved perturbative expansion (CIPT), and differences in the treatment and inclusion of non-perturbative corrections, result in systematic differences in the central values of $\alpha_s(M_\tau)$, ranging from 0.316 to 0.344, as summarised in Fig. 4. Results based on FOPT turn out to be systematically lower than those using CIPT—a trend being known for quite some time, and being actively disputed in the literature, but not finally being solved.

The results shown in Fig. 4, within their assigned total uncertainties, are partly incompatible with each other. This is especially true if considering that they are based on the same data sets. The main reason for these discrepancies is the usage of either the FOPT [45, 46, 48, 50] (marked *Beneke, Caprini, Maltmann* and *Narison*) or the CIPT [47, 49] (marked *Davier* and *Menke*) perturbative expansions. Only the result of Baikov et al. [18] averages between these two expansions, and assigns an overall error which includes the difference between these two.

---

[5]According to (1) and (2), in leading order, $\Delta\alpha_s(Q^2)/\alpha_s(Q^2) \sim \alpha_s(Q^2)$. Therefore, since $\alpha_s(Q^2)$ decreases by about a factor of 3 when running from $Q^2 = M_\tau^2$ to M_Z^2, the relative error of α_s also decreases by about a factor of 3.

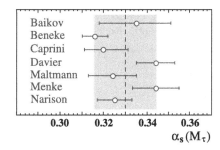

Fig. 4 Determinations of α_s from hadronic τ lepton decays [18, 45–50]. The results are all based on the same experimental data and on perturbative QCD predictions to $\mathcal{O}(\alpha_s^4)$, however vary in preference or range of the perturbative expansion and inclusion and treatment of non-perturbative corrections (see text). The *vertical line* and *shaded band* show the average value and uncertainty used as overall result from τ decays in this review

In view of these differences and for the sake of including the apparent span between different perturbative expansions in the overall error, the range shown as shaded band in Fig. 4 and the corresponding central value is taken as the final result from τ-decays, leading to

$$\alpha_s(M_\tau) = 0.330 \pm 0.014,$$

where the error is dominated by the theoretical uncertainty of the perturbative expansion. Running this value to the Z^0 rest mass of 91.2 GeV using the 4-loop solution of the β-function (see (6)) with 3-loop matching at the heavy quark pole masses $M_c = 1.5$ GeV and $M_b = 4.7$ GeV, results in

$$\alpha_s(M_{Z^0}) = 0.1197 \pm 0.0016.$$

This value will be included in determining the world average of $\alpha_s(M_{Z^0})$ as described in Sect. 4.

3.2 α_s from heavy quarkonia

Heavy Quarkonia, i.e. meson states consisting of a heavy (charm- or bottom-) quark and antiquark, are a classical testing ground for QCD, see e.g. [54]. Masses, mass splittings between various states, and decay rates are observables which can be measured quite accurately, and which can be predicted by QCD based on both perturbation theory and on lattice calculations.

3.2.1 α_s from radiative Υ decays

Bound states of a bottom quark and antiquark are potentially very sensitive to the value of α_s because the hadronic decay proceeds via three gluons, $\Upsilon \to ggg \to$ hadrons. The lowest-order QCD term (i.e. P_l in (8)) for the hadronic Υ decay width already contains α_s to the 3rd power. The situation is more complicated, however, due to relativistic corrections and to the unknown wave function of the Υ at the origin.

The wave function and relativistic corrections largely cancel out in ratios of decay widths like

$$R_\gamma = \frac{\Gamma(\Upsilon \to \gamma gg)}{\Gamma(\Upsilon \to ggg)}$$

which therefore are the classical observables for precise determinations of α_s.

In [55], recent CLEO data [56] are used to determine α_s from radiative decays of the $\Upsilon(1S)$. The theoretical predictions include QCD up to NLO ($\mathcal{O}(\alpha_s^3)$). They are based on recent estimates of colour octet operators and avoid any model dependences. The value obtained from this study is

$$\alpha_s(M_{Z^0}) = 0.119^{+0.006}_{-0.005}.$$

It is compatible with previous results from similar studies, see e.g. [10, 39] and references quoted therein. It will be included in the calculation of the new world average of $\alpha_s(M_{Z^0})$.

3.2.2 α_s from lattice QCD

Determinations of α_s based on lattice QCD calculations have become increasingly inclusive and precise in the past, including light quarks (u, d and s) in the vacuum polarisation and incorporating finer lattice spacing.

In a recent study by the HPQCD collaboration [37], the QCD parameters—the bare coupling constant and the bare quark masses—are tuned to reproduce the measured $\Upsilon'–\Upsilon$ meson mass difference. The u, d and s quark masses are adjusted to give correct values of various light meson masses. With these parameters set, there are no other free physical parameters, and the simulation is used to provide accurate QCD predictions. Non-perturbative values of several short-distance quantities are computed and compared to respective perturbative calculations which are given in NNLO perturbation theory. From a fit to 22 short-distance quantities, the value of

$$\alpha_s(M_{Z^0}) = 0.1183 \pm 0.0008$$

is finally obtained. The total error includes finite lattice spacing, finite lattice volume, perturbative and extrapolation uncertainties. This result will be an important ingredient of the new world average determined in Sect. 4.

3.3 α_s from deep inelastic scattering

Measurements of scaling violations in deep inelastic lepton–nucleon scattering belong to the earliest methods used to determine α_s. The first significant determinations of α_s, being based on perturbative QCD prediction in NLO, date back to 1979 [57].

Today, a large number of results is available from fixed target reactions using beams of electrons, muons or neutrinos, in the Q^2 range up to $\mathcal{O}$ (100 GeV^2). With the advent of the electron-proton collider HERA the Q^2 range has been extended by up to two orders of magnitude. In addition to scaling violations of structure functions, α_s is also determined from moments of structure functions, from QCD sum rules and—similar as in e$^+$e$^-$ annihilation—from hadronic jet production and event shapes. Improved QCD predictions as well as new experimental studies provided new results from a combined study of world data on structure functions, and from jet production at HERA.

3.3.1 α_s from world data on non-singlet structure functions

Perturbative predictions of physical processes in lepton-nucleon and in hadron–hadron collisions depend on quark- and gluon-densities in the nucleon. Assuming factorisation between short-distance, hard scattering processes which can be calculated using QCD perturbation theory, and low-energy or long-range processes which are not accessible by perturbative methods, such cross sections are parametrised by a set of structure functions F_i ($i = 1, 2, 3$) which are also interpreted as sums of parton (quark, antiquark and gluon) densities in the nucleon. While perturbative QCD cannot predict the functional form of parton densities and structure functions, their energy evolution is described by the so-called DGLAP equations [58–60].

A study [61] of the available world data on deep inelastic lepton–proton and lepton–deuteron scattering provided a determination of the valence quark parton densities and of α_s in wide ranges of the Bjorken scaling variable x and Q^2. In the non-singlet case, where heavy flavour effects are negligibly small, the analysis is extended to QCD in $\mathcal{O}(\alpha_\mathrm{s}^3)$ perturbative expansion.

The determination of α_s to this level results in

$$\alpha_\mathrm{s}(M_{Z^0}) = 0.1142 \pm 0.0023,$$

where the total error includes a theoretical uncertainty of ± 0.0008 which is taken from the difference between the N3LO and the NNLO result. This value will be included in the determination of the world average of $\alpha_\mathrm{s}(M_{Z^0})$.

As it appears, fits of α_s in determinations of parton density functions from deep inelastic scattering processes alone result in somewhat smaller values than those which also include hadron collider data, and they are systematically smaller than the world average value of $\alpha_\mathrm{s}(M_{Z^0})$, see below. For instance, in [62] which includes deep inelastic scattering and hadron collider data, a value of $\alpha_\mathrm{s}(M_{Z^0}) = 0.1171 \pm 0.0014_{\mathrm{exp.}}$ is obtained, with an additional theoretical uncertainty estimated to be smaller than ± 0.002.

3.3.2 α_s from jet production in deep inelastic scattering processes

Measurements of α_s from jet production in deep inelastic lepton–nucleon scattering at the HERA collider have been and continue to be an active field of research. Inclusive as well as differential jet production rates were studied in the energy range of $Q^2 \sim 10$ up to 15000 GeV^2, based on similar jet definitions and algorithms as used in e$^+$e$^-$ annihilation.

In a recent summary and combination [63] of precision measurements at HERA, values of α_s where determined from fits of NLO QCD predictions to data of inclusive jet cross sections in neutral current deep inelastic scattering at high Q^2 [64–66]. The overall combined result,

$$\alpha_\mathrm{s}(M_{Z^0}) = 0.1198 \pm 0.0032,$$

has a reduced theoretical uncertainty of ± 0.0026 (added in quadrature to the experimental error of ± 0.0019) compared to previous combinations, due to carefully selected ranges of data in Q^2 ranges where theoretical uncertainties are minimal [63]. This combined result will be included in the determination of the world average of $\alpha_\mathrm{s}(M_{Z^0})$.

3.4 α_s from hadronic event shapes and jet production in e$^+$e$^-$ annihilation

Observables parameterising hadronic event shapes and jet production rates are the classical inputs for α_s studies in e$^+$e$^-$ annihilation. The measurements summarised in previous reviews [10, 39, 67, 68] were based on QCD predictions in NLO, which partly included summation of next-to-leading logarithms (NLLA) to all orders. As one of the most notable and long awaited theoretical improvements, complete NNLO predictions became available recently [16, 17], which are also matched with leading and next-to-leading logarithms resummed to all orders [69] (NNLO + NLLA).

The advancement in theoretical descriptions was instantly used to determine α_s from data of previous e$^+$e$^-$ annihilation experiments, from the PETRA and the LEP colliders which operated from 1979 to 1986 and from 1989 to 2000, respectively. The usage of data of past experiments demonstrates the need to preserve data as well as reconstruction-, simulation- and analysis-software for a time-span significantly exceeding the usual ~5-year period of post-data taking analysis.

A re-analysis [70, 71] of the ALEPH data from LEP, in the c.m. energy range from 90 to 206 GeV, based on six event shape and jet production observables, results in

$$\alpha_\mathrm{s}(M_{Z^0}) = 0.1224 \pm 0.0039.$$

The total error contains an experimental uncertainty of 0.0013 and is dominated by a theoretical uncertainty, mainly from hadronisation and from renormalisation scale dependences, of 0.0037. This result is obtained using NNLO + NLLA QCD predictions; in NNLO alone, the central value is slightly higher (0.1240) and the total error is slightly smaller (0.0032). NNLA terms, although they should provide a more complete perturbation series, tend to introduce somewhat larger scale uncertainties [70, 71].

Similar results are available from a re-analysis of data from the JADE experiment at PETRA [72], from six event shape and jet observables at six c.m. energies in the c.m. energy range from 14 to 46 GeV:

$$\alpha_s(M_{Z^0}) = 0.1172 \pm 0.0051.$$

The total error contains an experimental uncertainty of 0.0020 and a theoretical uncertainty of 0.0046.[6] Also this result is obtained using QCD predictions in NNLO+NLLA; the value for NNLO alone is $\alpha_s(M_{Z^0}) = 0.1212 \pm 0.0060$.

Both the NNLO+NLLA results from ALEPH and from JADE data are retained for the determination of the new world average value of $\alpha_s(M_{Z^0})$ in this review. Because they are based on data at different c.m. energy ranges and from two independent experiments, they add valuable and independent information not only on the world average, but also on the experimental verification of the running of α_s. These two results of $\alpha_s(M_{Z^0})$ are included in Fig. 6; the respective values of $\alpha_s(Q)$, obtained at different values of c.m. energies, are displayed in Fig. 5.

Recently, the event shape observable thrust [73, 74] was also studied using methods of effective field theory [75]. Starting from a factorisation theorem in soft–collinear effective theory, the leading, next-to-leading, next-to-next-to-leading and the next-to-next-to-next-to-leading logarithmic terms (N3LL) of the thrust distribution are determined and are resummed to infinite order. This is two orders higher than previously known resummation terms. The N3LL terms are matched with the existing NNLO fixed-order results, and the resulting predictions are applied to the LEP thrust data.

The resulting value of α_s is $\alpha_s(M_{Z^0}) = 0.1172 \pm 0.0021$, whereby the error includes a theoretical uncertainty of ± 0.0017. Although this is formally one of the smallest errors quoted on measurements of $\alpha_s(M_{Z^0})$, this result is not explicitly included in the world average calculated below. The reason for this decision is two-fold: first, the LEP thrust data are already included in the re-analysis of ALEPH data described above. Second, this analysis based on effective field theory, although being a highly interesting, alternative approach to obtain and include higher than NNLO perturbative contributions, is not yet in a state of comparable reliability because it is based on one event shape observable only, and therefore misses an important verification of potential systematic uncertainties.

---

[6]At smaller c.m. energies, hadronisation but also perturbative uncertainties are larger than at LEP.

3.5 α_s from electroweak precision data

The determination of α_s from totally inclusive observables, like the hadronic width of the τ lepton discussed above, or the total hadronic decay width of the Z^0 boson, are of utmost importance because they lack many sources of systematic uncertainties, experimental as well as theoretical, which differential distributions like event shapes or jet rates suffer from. In this sense, the ratio of the hadronic to the leptonic partial decay width, $R_Z = \Gamma(Z^0 \to \text{hadrons})/\Gamma(Z^0 \to e^+e^-)$, is a "gold plated" observable, and fits of α_s and other quantities from precision electroweak measurements from e^+e^- annihilation and other processes offer excellent prospects not being plagued by hadronisation and other systematic uncertainties, see e.g. [76].

Since 1994, the QCD correction to R_Z is known in NNLO QCD [77, 78]. The measured value from LEP, $R_Z = 20.767 \pm 0.025$ [79], results in $\alpha_s(M_{Z^0}) = 0.1226 \pm 0.0038$, where the error is experimental. An additional theoretical uncertainty was estimated [10] as $^{+0.0043}_{-0.0005}$.

As already mentioned, the full N3LO prediction of R_Z, i.e. in $\mathcal{O}(\alpha_s^4)$ perturbative expansion, is now available [18]. The negative $\mathcal{O}(\alpha_s^4)$ term results in an increase of $\alpha_s(M_{Z^0})$ by 0.0005, such that the actual result from the measured value of R_Z is: $\alpha_s(M_{Z^0}) = 0.1231 \pm 0.0038$. Defining the remaining theoretical uncertainty as the difference between the NNLO and the N3LO result, the theory error would not visibly contribute any more, given the current size of the experimental error on $\alpha_s(M_{Z^0})$ of ± 0.0038.

A more precise value of $\alpha_s(M_{Z^0})$ can be obtained from general fits to all existing electro-weak precision data, using data from the LEP and the SLC e^+e^- colliders as well as measurements of the top-quark mass and limits on the Higgs boson mass from Tevatron and LEP. Such global fits result in values of $\alpha_s(M_{Z^0})$ with reduced experimental errors. These values, however, were consistently smaller than (but still compatible with) the ones obtained from R_Z alone, see e.g. [79].

A recent revision of the global fit to electroweak precision data [80], based on a new generic fitting package *Gfitter* [81], on the up-to-date QCD corrections in N3LO, on proper inclusion of the current limits from direct Higgs-searches at LEP and at the Tevatron and on other improved details, results in

$$\alpha_s(M_{Z^0}) = 0.1193^{+0.0028}_{-0.0027} \pm 0.0005,$$

where the first error is experimental and the second is theoretical, estimated by the difference of the results in NNLO and in N3LO QCD. This latter result will be included in the determination of the world average value of $\alpha_s(M_{Z^0})$.

4 The 2009 world average of $\alpha_s(M_{Z^0})$

The new results discussed in the previous sections are summarised in Table 1 and in Figs. 5 and 6. Since all of them are based on improved theoretical predictions and methods, and/or on improved data quality and statistics, they supersede and replace their respective precursor results which were summarised in a previous review [39]. While those previous results continue to be valid measurements, they are not discussed in this review again, and they will not be included in the determination of a combined world average values of $\alpha_s(M_{Z^0})$, according to the following reasons:

1. from each class of measurements, only the most advanced and complete analyses shall be included in the new world average;
2. older measurements *not* being complemented or superseded by the most recent results listed above, as e.g. results from sum rules, from singlet structure functions of deep inelastic scattering, and from jet production and b-quark production at hadron colliders, are not included because their relatively large overall uncertainties, in general, will not give them a sizable weight but will complicate the definition of the overall error of the combined value of $\alpha_s(M_{Z^0})$;
3. restricting the new world average to the most recent and most complete (i.e. precise) results allows one to examine the consistency between the newest and the previous generations of measurements and reviews.

4.1 Numerical procedure

The average $\overline{x}$ of a set of n different, uncorrelated measurements x_i of a particular quantity x with individual errors or uncertainties, Δx_i, is commonly defined using the method of *least squares* (see e.g. [9]): For x_i being independent and statistically distributed measures with a common

expectation value $\overline{x}$ but with different variances $(\Delta x_i)^2$, the weighted average is defined by

$$\overline{x} = \frac{\sum_{i=1}^n w_i x_i}{\sum_{i=1}^n w_i} \tag{14}$$

and the variance $(\Delta \overline{x})^2$ of $\overline{x}$ is minimised by choosing

$$(\Delta \overline{x})^2 = \frac{1}{\sum_{i=1}^n \frac{1}{(\Delta x_i)^2}}, \quad \text{i.e.} \quad w_i = \frac{1}{(\Delta x_i)^2}. \tag{15}$$

The quality of the average is defined by the χ^2 variable,

$$\chi^2 = \sum_{i=1}^n \frac{(x_i - \overline{x})^2}{(\Delta x_i)^2} \tag{16}$$

which is, for uncorrelated data, expected to be equal to the number of degrees of freedom, n_{df}:

$$\chi^2 = n_{df} = n - 1.$$

The results summarised in Table 1, however, are not independent of each other. They are, in the most general case, correlated to an unknown degree. While the statistical errors of the data and the experimental systematic uncertainties contained in the errors are independent and uncorrelated, the theoretical uncertainties are very likely to be (partly) correlated between different results, because they are all based on applying perturbative QCD, and similar methods to obtain estimates of theoretical uncertainties are being used.

For some observables, like e.g. the hadronic widths of the Z^0 boson and the τ lepton, the theoretical predictions and hence, their uncertainties, are known to be correlated by almost 100%. For other cases, like the results based on lattice QCD and those based on QCD perturbation theory, their theoretical uncertainties may not be correlated at all. In addition to the inherent lack of knowledge of theoretical

Table 1 Summary of recent measurements of $\alpha_s(M_{Z^0})$. All eight measurements are described in Sect. 3 and are included in determining the world average value of $\alpha_s(M_{Z^0})$. The rightmost two columns give the exclusive mean value of $\alpha_s(M_{Z^0})$ calculated *without* that particular measurement, and the number of standard deviations between this measurement and the respective exclusive mean, treating errors as described in Sect. 4. The inclusive average from *all* listed measurements gives $\alpha_s(M_{Z^0}) = 0.11842 \pm 0.00067$

Process	Q [GeV]	$\alpha_s(Q)$	$\alpha_s(M_{Z^0})$	excl. mean $\alpha_s(M_{Z^0})$	std. dev.
τ-decays	1.78	0.330 ± 0.014	0.1197 ± 0.0016	0.11818 ± 0.00070	0.9
DIS [F_2]	2–170	–	0.1142 ± 0.0023	0.11876 ± 0.00123	1.7
DIS [e-p $\rightarrow$ jets]	6–100	–	0.1198 ± 0.0032	0.11836 ± 0.00069	0.4
$Q\overline{Q}$ states	7.5	0.1923 ± 0.0024	0.1183 ± 0.0008	0.11862 ± 0.00114	0.2
Υ decays	9.46	$0.184^{+0.015}_{-0.014}$	$0.119^{+0.006}_{-0.005}$	0.11841 ± 0.00070	0.1
e^+e^- [jets & shps]	14–44	–	0.1172 ± 0.0051	0.11844 ± 0.00076	0.2
e^+e^- [ew prec. data]	91.2	0.1193 ± 0.0028	0.1193 ± 0.0028	0.11837 ± 0.00076	0.3
e^+e^- [jets & shps]	91–208	–	0.1224 ± 0.0039	0.11831 ± 0.00091	1.0

correlations, estimates of theoretical uncertainties, in general, are performed in widely different ways, using different methods and different ranges of parameters.

The presence of correlations, if using the equations given above, is usually signalled by $\chi^2 < n_{df}$. Values of $\chi^2 > n_{df}$, in most practical cases, are a sign of possibly underestimated errors. In this review, both these cases are pragmatically handled in the following way:[7]

In the presence of correlated errors, described by a covariance matrix C, the optimal procedure to determine the average $\overline{x}$ of a sample of measurements x_i is to minimise the χ^2 function

$$\chi^2 = \sum_{i,j=1}^{n} (x_i - \overline{x})\big(C^{-1}\big)_{ij}(x_j - \overline{x}), \qquad (17)$$

which leads to

$$\overline{x} = \left(\sum_{ij}(C^{-1})_{ij}x_j\right)\left(\sum_{ij}(C^{-1})_{ij}\right)^{-1} \qquad (18)$$

and

$$\Delta\overline{x}^2 = \left(\sum_{ij}(C^{-1})_{ij}\right)^{-1}. \qquad (19)$$

The choice of $C_{ii} = (\Delta x_i)^2$ and $C_{ij} = 0$ for $i \neq j$ retains the uncorrelated case given above. In the presence of correlations, however, the resulting χ^2 will be less than $n_{df} = n - 1$. In cases where correlations between a particular pair of measurements i and j are known or expected, the corresponding non-diagonal matrix elements C_{ij} and C_{ji} are set to $\rho\Delta x_i \Delta x_j$, where ρ is the respective correlation coefficient ranging between 0 (uncorrelated) and 1 (100% correlation).

If the resulting χ^2 is still smaller than n_{df}, the method proposed in [82] will be applied: an unknown additional, *common* degree of a correlation f is introduced between all measurements, by choosing $C_{ij} = f \times \Delta x_i \times \Delta x_j$ for $i \neq j$, and f is adjusted such that $\chi^2 = n_{df}$.

In cases where the assumption of uncorrelated errors results in $\chi^2 > n_{df}$, and without knowledge about which of the errors Δx_i are possibly underestimated, all individual errors are scaled up by a common factor g such that the resulting value of χ^2/n_{df}, using the definition for uncorrelated errors, will equal unity.

Note that both for values of $f > 0$ or $g > 1$, $\Delta\overline{x}$ increases, compared to the uncorrelated ($f = 0$ and $g = 1$) case.

---

[7]Since most measurements and their respective experimental and theoretical errors are defined and estimated in different ways, in this review, as already done previously, only the *total* uncertainties are considered, and no attempt is made to consistently separate experimental and theoretical errors.

4.2 Determination of the world average

The eight different determinations of $\alpha_s(Q^2)$ summarised and discussed in the previous sections are listed in Table 1 and are displayed in Fig. 5. The energy dependence of these results exactly follows the expectation of the QCD prediction of the running coupling. It is therefore straightforward to extrapolate all measurements of $\alpha_s(Q^2)$ to the common scale of M_Z, using the procedures and equations given in Sect. 2. The corresponding values of $\alpha_s(M_{Z^0})$ are listed in Table 1 and displayed in Fig. 6. Applying (14), (15) and

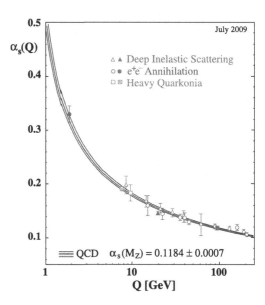

Fig. 5 Summary of measurements of α_s as a function of the respective energy scale Q. The *curves* are QCD predictions for the combined world average value of $\alpha_s(M_{Z^0})$, in 4-loop approximation and using 3-loop threshold matching at the heavy quark pole masses $M_c = 1.5$ GeV and $M_b = 4.7$ GeV. *Full symbols* are results based on N3LO QCD, *open circles* are based on NNLO, *open triangles* and *squares* on NLO QCD. The *cross-filled square* is based on lattice QCD. The *filled triangle* at $Q = 20$ GeV (from DIS structure functions) is calculated from the original result which includes data in the energy range from $Q = 2$ to 170 GeV

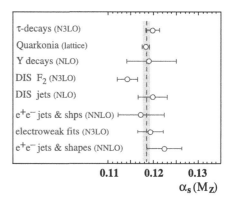

Fig. 6 Summary of measurements of $\alpha_s(M_{Z^0})$. The *vertical line* and *shaded band* mark the final world average value of $\alpha_s(M_{Z^0}) = 0.1184 \pm 0.0007$ determined from these measurements

(16) to this set of measurements, assuming that the errors are not correlated, results in an average value of $\alpha_s(M_{Z^0}) = 0.11842 \pm 0.00063$ with $\chi^2/n_{df} = 5.4/7$.

The fact that $\chi^2 < n_{df}$ signals a possible correlation between all or subsets of the eight input results. Assuming an overall correlation factor f and demanding that $\chi^2 = n_{df} = 7$ requires $f = 0.23$, inflating the overall error from 0.00063 to 0.00089.

In fact, there are two pairs of results which are known to be largely correlated:

- the two results from $e^+ e^-$ event shapes based on the data from JADE and from ALEPH use the same theoretical predictions and similar hadronisation models to correct these predictions for the transitions of quarks and gluons to hadrons. While the experimental errors are uncorrelated, the theoretical uncertainties may be assumed to be correlated to 100%. The latter account for about 2/3 to 3/4 of the total errors. An appropriate choice of correlation factor between the two may then be $f = 0.67$.
- the QCD predictions for the hadronic widths of the τ-lepton and the Z^0 boson are essentially identical, so the respective results on α_s are correlated, too. The values and total errors of $\alpha_s(M_{Z^0})$ from τ decays must therefore be correlated to a large extend, too. In this case, however, the error of one measurement is almost entirely determined by the experimental error (Z^0-decays), while the other, from τ-decays, is mostly theoretical. A suitable choice of the correlation factor between both these results may thus be $f = 0.5$.

Inserting these two pairs of correlations into the error matrix C, the χ^2/n_{df} of the averaging procedure results in 6.8/7, and the overall error on the (unchanged) central value of $\alpha_s(M_{Z^0})$ changes from 0.00063 to 0.00067. Therefore the new world average value of $\alpha_s(M_{Z^0})$ is defined to be

$$\boxed{\alpha_s(M_{Z^0}) = 0.1184 \pm 0.0007.}$$

For seven out of the eight measurements of $\alpha_s(M_{Z^0})$, the average value of 0.1184 is within one standard deviation of their assigned errors. One of the measurements, from structure functions [61], deviates from the mean value by more than one standard deviation, see Fig. 6.

The mean value of $\alpha_s(M_{Z^0})$ is potentially dominated by the α_s result with the smallest overall assigned uncertainty, which is the one based on lattice QCD [37]. In order to verify this degree of dominance on the average result and its error, and to test the compatibility of each of the measurements with the others, exclusive averages, leaving out one of the 8 measurements at a time, are calculated. These are presented in the 5th column of Table 1, together with the

corresponding number of standard deviations[8] between the exclusive mean and the respective single measurement.

As can be seen, the values of exclusive means vary only between a minimum of 0.11818 and a maximum 0.11876. Note that in the case of these exclusive means and according to the "rules" of calculating their overall errors, in four out of the eight cases small error scaling factors of $g = 1.06 \ldots 1.08$ had to be applied, while in the other cases, overall correlation factors of about 0.1, and in one case of 0.7, had to be applied to assure $\chi^2/n_{df} = 1$. Most notably, the average value $\alpha_s(M_{Z^0})$ changes to $\alpha_s(M_{Z^0}) = 0.1186 \pm 0.0011$ when omitting the result from lattice QCD.

5 Summary and discussion

In this review, new results and measurements of α_s are summarised, and the world average value of $\alpha_s(M_{Z^0})$, as previously given in [9, 10, 39], is updated. Based on eight recent measurements, which partly use new and improved N3LO, NNLO and lattice QCD predictions, the new average value is

$$\alpha_s(M_{Z^0}) = 0.1184 \pm 0.0007,$$

which corresponds to

$$\Lambda_{\overline{MS}}^{(5)} = (213 \pm 9)\ \text{MeV}.$$

This result is consistent with the one obtained in the previous review three years ago [39], which was $\alpha_s(M_{Z^0}) = 0.1189 \pm 0.0010$. The previous and the actual world average have been obtained from a non-overlapping set of single results; their agreement therefore demonstrates a large degree of compatibility between the old and the new, largely improved set of measurements and theoretical methods.

The individual measurements, as listed in Table 1 and displayed in Fig. 6, show a very satisfactory agreement with each other and with the overall average: only one out of eight measurements exceeds a deviation from the average by more than one standard deviation, and the largest deviation between any two out of the eight results, namely the ones from τ decays and from structure functions, amounts to 2 standard deviations.[9]

There remains, however, an apparent and long-standing systematic difference: results from structure functions favour smaller values of $\alpha_s(M_{Z^0})$ than most of the others, i.e. those from $e^+ e^-$ annihilations, from τ decays, but also those from jet production in deep inelastic scattering. This issue apparently remains to be true, although almost all of the new

---

[8] The number of standard deviations is defined as the square-root of the value of χ^2.

[9] Assuming their assigned total errors to be fully uncorrelated.

results are based on significantly improved QCD predictions, up to N3LO for structure functions, τ and Z^0 hadronic widths, and NNLO for e^+e^- event shapes.

The reliability of "measurements" of α_s based on "experiments" on the lattice have gradually improved over the years, too. Including vacuum polarisation of three light quark flavours and extended means to understand and correct for finite lattice spacing and volume effects, the overall error of these results significantly decreased over time, while the value of $\alpha_s(M_{Z^0})$ gradually approached the world average. Lattice results today quote the smallest overall error on $\alpha_s(M_{Z^0})$; it is, however, ensuring to see and note that the world average *without* lattice results is only marginally different, while the small size of the total uncertainty on the world average is, naturally, largely influenced by the lattice result.

In order to demonstrate the agreement of measurements with the specific energy dependence of α_s predicted by QCD, in Fig. 5 the recent measurements of α_s are shown as a function of the energy scale Q. For those results which are based on several α_s determinations at different values of energy scales Q, the individual values of $\alpha_s(Q)$ are displayed. For the value from structure functions such a breakup is not possible; instead, the corresponding result derived for a typical energy scale of $Q = 20$ GeV is displayed.

The measurements significantly prove the validity of the concept of asymptotic freedom; they are in perfect agreement with the QCD prediction of the running coupling. This is further corroborated by Fig. 7, where a selected sample of the measurements is plotted, now as a function of $1/\log Q$, in order to demonstrate the data reproducing the specific logarithmic shape of the running as predicted by QCD, signalling that indeed $\alpha_s(Q) \to 0$ for $Q \to \infty$.

What are the future prospects of measurements of α_s? With the given degree of data and theory precision, further improvements will be difficult and may take quite some time. Experimentally, a linear e^+e^- collider, especially if run in the "Giga-Z" mode, has the potential to decrease the dominating experimental error of $\alpha_s(M_{Z^0})$ from the measurement of R_Z, down to and below its theoretical uncertainty which currently is assumed to be, in N3LO, ± 0.0005.

While it is unlikely that QCD perturbation theory will improve to yet one order higher than the existing N3LO or NNLO predictions, improvements are likely, and actually are very necessary, for QCD predictions of jet production in deep inelastic scattering and in hadron collisions, where calculations currently are limited to NLO. The precision of QCD tests, but also the sensitivity for observing new physics signals at the LHC, will largely depend on a further advancement of QCD predictions for hadron collisions.

Last but not least, further developments of non- perturbative methods are mandatory to bridge the gap between quarks and gluons and their final states, hadrons. They may

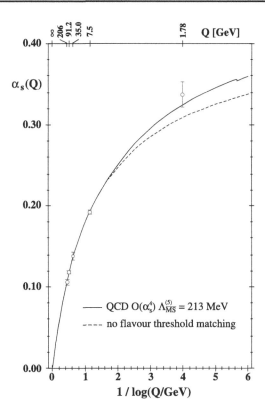

Fig. 7 Selected measurements of α_s, as a function of the inverse logarithm of the energy scale Q, in order to demonstrate concordance with Asymptotic Freedom. The *full line* is the QCD prediction in 4-loop approximation with 3-loop threshold matching at the heavy quark pole masses. The *dashed line* indicates extrapolation of the 5-flavour prediction without threshold matching

in fact shed more light into the systematic differences between some classes of measurements as discussed above.

Future improvement of theoretical predictions and models require the conservation of data and of reconstruction and simulation code of current and past experiments; especially in the case of deep inelastic scattering data, reapplication of improved predictions and models carry a large potential for future advancements in this field.

References

1. H. Fritzsch, M. Gell-Mann, H. Leutwyler, Phys. Lett. B **47**, 365 (1973)
2. D.J. Gross, F. Wilczek, Phys. Rev. Lett. **30**, 1343 (1973)
3. D.J. Gross, F. Wilczek, Phys. Rev. D **8**, 3633 (1973)
4. H.D. Politzer, Phys. Rev. Lett. **30**, 1346 (1973)
5. R.K. Ellis, W.J. Stirling, B.R. Webber, *QCD and Collider Physics* (Cambridge University Press, Cambridge, 1996)
6. J.C. Collins, *Renormalization* (Cambridge University Press, Cambridge, 1984)
7. F.J. Yndurain, *The Theory of Quark and Gluon Interactions* (Springer, Berlin, 1999)
8. G. Dissertori, I.G. Knowles, M. Schmelling, *High Energy Experiments and Theory*. International Series of Monographs on Physics, vol. 115 (Clarendon, Oxford, 2003), p. 538

9. C. Amsler et al. (PDG), Phys. Lett. B **667** (2008)
10. S. Bethke, J. Phys. G **26**, R27 (2000). hep-ex/0004021
11. T. van Ritbergen, J.A.M. Vermaseren, S.A. Larin, Phys. Lett. B **400**, 379 (1997)
12. M. Czakon, Nucl. Phys. B **710**, 485 (2005). hep-ph/0411261
13. K.G. Chetyrkin, B.A. Kniehl, M. Steinhauser, Phys. Rev. Lett. **79**, 2184 (1997)
14. W. Bernreuther, W. Wetzel, Nucl. Phys. B **197**, 128 (1982)
15. S.A. Larin, T. van Ritbergen, J.A.M. Vermaseren, Nucl. Phys. B **438**, 278 (1995)
16. A. Gehrmann-de Ridder et al., J. High Energy Phys. **0712**, 094 (2007). arXiv:0711.4711 [hep-ph]
17. S. Weinzierl, Phys. Rev. Lett. **101**, 162001 (2008). arXiv:0807.3241 [hep-ph]
18. P.A. Baikov, K.G. Chetyrkin, J.H. Kühn, Phys. Rev. Lett. **101**, 012002 (2008). arXiv:0801.1821 [hep-ph]
19. S. Catani, L. Trentadue, G. Turnock, B.R. Webber, Nucl. Phys. B **407**, 3 (1993)
20. P.M. Stevenson, Phys. Rev. D **23**, 2916 (1981)
21. G. Grunberg, Phys. Rev. D **29**, 2315 (1984)
22. S.J. Brodsky, G.P. Lepage, P.B. Mackenzie, Phys. Rev. D **28**, 228 (1983)
23. S. Bethke, Z. Phys. C **43**, 331 (1989)
24. W.A. Bardeen et al., Phys. Rev. D **18**, 3998 (1978)
25. T. Sjostrand, Comput. Phys. Commun. **27**, 243 (1982)
26. T. Sjostrand, S. Mrenna, P. Skands, Comput. Phys. Commun. **178**, 852 (2008). arXiv:0710.3820 [hep-ph]
27. G. Marchesini, B.R. Webber, Nucl. Phys. B **238**, 1 (1984)
28. G. Corcella et al., J. High Energy Phys. **0101**, 010 (2001). hep-ph/0011363
29. Yu.L. Dokshitzer, B.R. Webber, Phys. Lett. B **352**, 451 (1995)
30. Yu.L. Dokshitzer, G. Marchesini, B.R. Webber, Nucl. Phys. B **469**, 93 (1996)
31. Yu.L. Dokshitzer, B.R. Webber, Phys. Lett. B **404**, 321 (1997)
32. S. Catani, B.R. Webber, Phys. Lett. B **427**, 377 (1998)
33. Yu.L. Dokshitzer, A. Lucenti, G. Marchesini, G.P. Salam, Nucl. Phys. B **511**, 296 (1998)
34. Yu.L. Dokshitzer, A. Lucenti, G. Marchesini, G.P. Salam, J. High Energy Phys. **05**, 003 (1998)
35. P. Weisz, Nucl. Phys. B (Proc. Suppl.) **47**, 71 (1996). hep-lat/9511017
36. S. Durr et al., Science **322**, 1224 (2008). arxiv:0906.3599 [hep-lat]
37. C.T.H. Davies et al. (HPQCD Collaboration), Phys. Rev. D **78**, 114507 (2008). arXiv:0807.1687 [hep-lat]
38. G. Altarelli, Ann. Rev. Nucl. Part. Sci. **39**, 357 (1989)
39. S. Bethke, Prog. Part. Nucl. Phys. **58**, 351 (2007). hep-ex/0606035
40. E. Braaten, S. Narison, A. Pich, Nucl. Phys. B **373**, 581 (1992)
41. W. Marciano, A. Sirlin, Phys. Rev. Lett. **61**, 1815 (1988)
42. E. Braaten, C.S. Li, Phys. Rev. D **42**, 3888 (1990)
43. M.A. Shifman, L.A. Vainshtein, V.I. Zakharov, Nucl. Phys. B **147**, 385 (1979)
44. M. Davier, A. Höcker, Z. Zhang, Rev. Mod. Phys. **78**, 1043 (2006). hep-ph/0507078
45. M. Beneke, M. Jamin, M. Beneke, J. High Energy Phys. **0809**, 044 (2008). arXiv:0806.3156 [hep-ph]
46. I. Caprini, J. Fischer, arXiv:0906.5211 [hep-ph]; Eur. Phys. J. C. doi:10.1140/epjc/s10052-009-1142-8

47. M. Davier et al., Eur. Phys. J. C **56**, 305 (2008). arXiv:0803.0979 [hep-ph]
48. K. Maltman, T. Yavin, Phys. Rev. D **78**, 094020 (2008). arXiv:0807.0650 [hep-ph]
49. S. Menke, arXiv:0904.1796 [hep-ph]
50. S. Narison, Phys. Lett. B **673**, 30 (2009). arXiv:0901.3823 [hep-ph]
51. K. Ackerstaff et al. (OPAL Collaboration), Eur. Phys. J. C **7**, 571 (1999). hep-ex/9808019
52. S. Schael et al. (ALEPH Collaboration), Phys. Rep. **421**, 191 (2005). arXiv:hep-ex/0506072v1
53. B. Aubert et al. (BABAR Collaboration), Phys. Rev. Lett. **100** (2008). arXiv:0707.2981 [hep-ex]
54. N. Brambilla et al. (CERN-2005-005), hep-ph/0412158
55. N. Brambilla et al., Phys. Rev. D **75**, 074014 (2007). hep-ph/0702079
56. D. Besson et al. (CLEO Collaboration), Phys. Rev. **74**, 012003 (2006). hep-ex/0512061
57. A. Gonzales-Arroyo, C. Lopez, F.J. Yndurain, Nucl. Phys. B **153**, 161 (1979)
58. V.N. Gribov, L.N. Lipatov, Sov. J. Nucl. Phys. **15**, 438 (1972)
59. Yu.L. Dokshitzer, Sov. Phys. JETP **46**, 641 (1977)
60. G. Altarelli, G. Parisi, Nucl. Phys. B **126**, 298 (1977)
61. J. Blümlein, H. Böttcher, A. Guffanti, Nucl. Phys. B **774**, 182 (2007). hep-ph/0607200
62. A.D. Martin et al., Phys. Lett. B **652**, 292 (2007). arXiv:0706.0459 [hep-ph], arXiv:0905.3531 [hep-ph]
63. C. Glasman, J. Phys. Conf. Ser. **110**, 022013 (2008). arXiv:0709.4426
64. S. Chekanov et al. (ZEUS Collaboration), Phys. Lett. B **649**, 12 (2007). hep-ex/0701039
65. A. Aktas et al. (H1 Collaboration), Phys. Lett. B **653**, 134 (2007). arXiv:0706.3722 [hep-ex]
66. F.D. Aaron et al. (H1 Collaboration), arXiv:0904.3870 [hep-ex]
67. O. Biebel, Phys. Rep. **340**, 165 (2001)
68. S. Kluth, Rep. Prog. Phys. **69**, 1771 (2006). hep-ex/0603011
69. T. Gehrmann, G. Luisoni, H. Stenzel, Phys. Lett. B **664**, 265 (2008). arXiv:0803.0695
70. G. Dissertori et al., J. High Energy Phys. **0802**, 040 (2008). arXiv:0712.0327 [hep-ph]
71. G. Dissertori et al., arXiv:0906.3436 [hep-ph]
72. S. Bethke et al., arXiv:0810.1389 [hep-ex]; Eur. Phys. J. C. doi:10.1140/epjc/s10052-009-1149-1
73. S. Brandt et al., Phys. Lett. **12**, 57 (1964)
74. E. Farhi, Phys. Rev. Lett. **39**, 1587 (1977)
75. T. Becher, M.D. Scheartz, J. High Energy Phys. **0807**, 034 (2008). arXiv:0803.0342 [hep-ph]
76. K. Hagiwara, D. Haidt, S. Matsumoto, Eur. Phys. J. C **2**, 95 (1998). hep-ph/9706331
77. S.A. Larin, T. van Ritbergen, J.A.M. Vermaseren, Phys. Lett. B **320**, 159 (1994)
78. K.G. Chetyrkin, O.V. Tarasov, Phys. Lett. B **327**, 114 (1994)
79. The LEP Collaborations ALEPH, DELPHI, L3 and OPAL, hep-ex/0509008
80. H. Flächer et al., Eur. Phys. J. C **60**, 543 (2009). arXiv:0811.0009 [hep-ph]
81. http://cern.ch/Gfitter
82. M. Schmelling, Phys. Scr. **51**, 676 (1995)

Printed in the United States
By Bookmasters